Solutions for High-Touch Communications in a High-Tech World

Michael A. Brown Sr.
Florida International University, USA

A volume in the Advances in Human and Social Aspects of Technology (AHSAT) Book Series

IGI GLOBAL
DISSEMINATOR OF KNOWLEDGE
www.igi-global.com

Published in the United States of America by
IGI Global
Information Science Reference (an imprint of IGI Global)
701 E. Chocolate Avenue
Hershey PA 17033
Tel: 717-533-8845
Fax: 717-533-8661
E-mail: cust@igi-global.com
Web site: http://www.igi-global.com

Library of Congress Cataloging-in-Publication Data

Names: Brown, Michael A., Sr., 1956- author.
Title: Solutions for high-touch communications in a high-tech world / by Michael A. Brown Sr.
Description: Hershey, PA : CyberTech Publishing, [2017]
Identifiers: LCCN 2016046357| ISBN 9781522518976 (hardcover) | ISBN 9781522518983 (ebook)
Subjects: LCSH: Interpersonal communication--Technological innovations--Social aspects. | Communication--Technological innovations. | Online social networks. | Digital communications.
Classification: LCC HM1166 .B76 2017 | DDC 302.23/1--dc23 LC record available at https://lccn.loc.gov/2016046357

This book is published in the IGI Global book series Advances in Human and Social Aspects of Technology (AHSAT) (ISSN: 2328-1316; eISSN: 2328-1324)

British Cataloguing in Publication Data
A Cataloguing in Publication record for this book is available from the British Library.

Advances in Human and Social Aspects of Technology (AHSAT) Book Series

ISSN:2328-1316
EISSN:2328-1324

Editor-in-Chief: Ashish Dwivedi, The University of Hull, UK

Mission

In recent years, the societal impact of technology has been noted as we become increasingly more connected and are presented with more digital tools and devices. With the popularity of digital devices such as cell phones and tablets, it is crucial to consider the implications of our digital dependence and the presence of technology in our everyday lives.

The **Advances in Human and Social Aspects of Technology (AHSAT) Book Series** seeks to explore the ways in which society and human beings have been affected by technology and how the technological revolution has changed the way we conduct our lives as well as our behavior. The AHSAT book series aims to publish the most cutting-edge research on human behavior and interaction with technology and the ways in which the digital age is changing society.

Coverage

- ICTs and social change
- Philosophy of technology
- Cyber Behavior
- Technology Adoption
- Cultural Influence of ICTs
- Technology Dependence
- Computer-Mediated Communication
- Gender and Technology
- ICTs and human empowerment
- Technology and Social Change

IGI Global is currently accepting manuscripts for publication within this series. To submit a proposal for a volume in this series, please contact our Acquisition Editors at Acquisitions@igi-global.com or visit: http://www.igi-global.com/publish/.

The Advances in Human and Social Aspects of Technology (AHSAT) Book Series (ISSN 2328-1316) is published by IGI Global, 701 E. Chocolate Avenue, Hershey, PA 17033-1240, USA, www.igi-global.com. This series is composed of titles available for purchase individually; each title is edited to be contextually exclusive from any other title within the series. For pricing and ordering information please visit http://www.igi-global.com/book-series/advances-human-social-aspects-technology/37145. Postmaster: Send all address changes to above address.

Titles in this Series

For a list of additional titles in this series, please visit:
http://www.igi-global.com/book-series/advances-educational-technologies-instructional-design/73678

Enriching Urban Spaces with Ambient Computing, the Internet of Things, and Smart...
Shin'ichi Konomi (University of Tokyo, Japan) and George Roussos (University of London, UK)
Engineering Science Reference • ©2017 • 323pp • H/C (ISBN: 9781522508274) • US $210.00

Handbook of Research on Individualism and Identity in the Globalized Digital Age
F. Sigmund Topor (Keio University, Japan)
Information Science Reference • ©2017 • 645pp • H/C (ISBN: 9781522505228) • US $295.00

Information Technology Integration for Socio-Economic Development
Titus Tossy (Mzumbe University, Tanzania)
Information Science Reference • ©2017 • 385pp • H/C (ISBN: 9781522505396) • US $200.00

Handbook of Research on Human-Computer Interfaces, Developments, and Applications
João Rodrigues (Univ. of Algarve, Portugal) Pedro Cardoso (Univ. of Algarve, Portugal) Jânio Monteiro (Univ. of Algarve, Portugal) and Mauro Figueiredo (Univ. of Algarve, Portugal)
Information Science Reference • ©2016 • 663pp • H/C (ISBN: 9781522504351) • US $330.00

Human Development and Interaction in the Age of Ubiquitous Technology
Hakikur Rahman (BRAC University, Bangladesh)
Information Science Reference • ©2016 • 384pp • H/C (ISBN: 9781522505563) • US $185.00

Examining the Evolution of Gaming and Its Impact on Social, Cultural, and Political...
Keri Duncan Valentine (West Virginia Univ., USA) and Lucas John Jensen (Georgia Southern Univ., USA)
Information Science Reference • ©2016 • 456pp • H/C (ISBN: 9781522502616) • US $190.00

Handbook of Research on Human Social Interaction in the Age of Mobile Devices
Xiaoge Xu (Botswana International University of Science and Technology, Botswana)
Information Science Reference • ©2016 • 548pp • H/C (ISBN: 9781522504696) • US $325.00

Defining Identity and the Changing Scope of Culture in the Digital Age
Alison Novak (Rowan University, USA) and Imaani Jamillah El-Burki (Lehigh University, USA)
Information Science Reference • ©2016 • 316pp • H/C (ISBN: 9781522502128) • US $185.00

For an enitre list of titles in this series, please visit:
http://www.igi-global.com/book-series/advances-human-social-aspects-technology/37145

IGI GLOBAL
DISSEMINATOR OF KNOWLEDGE

www.igi-global.com

701 East Chocolate Avenue, Hershey, PA 17033, USA
Tel: 717-533-8845 x100 • Fax: 717-533-8661
E-Mail: cust@igi-global.com • www.igi-global.com

Table of Contents

Section 3
Engaged Interaction

Foreword

SOCIAL MEDIA'S ROLE IN UNDERSTANDING AND OVERCOMING THE COMMUNICATION GAP

Successful communication is mandatory for successful business, educational institutions, and even families to attain shared goals and objectives. Strangely enough, the Internet offers some meaningful transparency, filling in gaps in communication that can provide depth of meaning to the individual or the organization. Social media can be used by any organization that wants to improve and deepen its communication process. Dr. Brown's book addresses solutions to common communication gaps in business organizations via use of social media.

Communication is an exchange of information between groups or individuals using mutually understood systems of symbols and semantics flavored with tones of behavior. In virtually any setting, meeting effective communication requires that the meaning intended by the speaker or sender is completely understood by the recipient. While social media often delivers an instant message, unlike spoken language, the message is documented, allowing for review, study, and alternative interpretations. The message, if crafted correctly, does not evaluate the receiver's intelligence or workplace positon. In our fast-paced multigenerational cultural universe it is sometime painful to see talented individuals of all generations passed by because they dismiss the importance of social media messages, or more commonly, lack the ability to adapt to the constantly evolving technologies of social media. Some communicators, because of their youth, have not been taught the nuances of communication delivered via social media. Communication is still very necessary to survive in life and in an individual's career pathway.

America is walking on razor blades. As a society we need to just glance behind our collective shoulder to 2008 and 2009, when hundreds of thousands of Baby Boomers had their career pathway, so carefully constructed, crumble when their positons were more than eliminated. They became unnecessary, easily replaced by technology. This book will help older generations understand the importance of social media in an entirely new and meaningful way.

Our Millennials are now in our workforce. Squeezed between Boomers and Gen X, they appear to be born with the know-how of cyber communication.

Millennials, the first generation to grow up with a screen in hand may have had more than their zeitgeist shaped by digital media; their brains function differently. It's true, EEG readings of brain activity by Nielsen NeuroFocus reports important differences in this cohort's engagement and retention of messages, compared to older consumers.

Yet while Millennials may engage better with more information, movement, chaos, or clutter, their digital connection robs this population of the necessary social interaction that defines our culture and ultimately our worldwide society. (Adapted from: www.nielsen.com/us/en/solutions/capabilities/consumer-neuroscience.html)

Rest assured, America and the rest of the world has not spawned a generation of workers less dedicated to productivity. And certainly all members of all generations have the intrinsic deep desire to communicate. For the last several years, my colleague Michael Brown, Ph.D. has dedicated his research to the "how" of this new communication tool. For over a decade, his diligent studies have explored and now explain social media and how it operates in parallel with interactions that are as intimate as face-to-face conversations.

Brown has proven to skeptics like me that social media and the tools of the Internet offer intensely personal and moving ways to communicate. *Solutions for High-Touch Communications in a High-Tech World* stands out because it is the culmination of qualitative and quantitative research into how our newest and possibly best solution to meaningful communication operates. Social media is plastic and can take shape and become an effective tool for learning and business. Correctly used, social media involves honest human social interaction. The challenge to the reader and leaders of the future is to take the basics offered within these pages and convert them to a personalized guidebook that can be applied to your life, regardless of age, work experience, or cultural situation. What a gift.

Millions worldwide may be working with new technologies, but how to *adapt* in a healthy way as human kind has done through the millennia has not been explored through the eyes of social media, which is now our dominate form of communication. Brown explores and then teaches how technology becomes a human tool for not only new efficiencies, but for building relationships, fostering emotions, and providing a strong communication bridge between world cultures.

There is an important subtext to this publication. *Technology is not our new religion.* Technology is not who we report to. We are not defined by technology. Technology is a very useful, albeit incredibly fast-moving tool. But it is a *human tool* and can be used by society to enhance communication in entirely new ways.

In 1961 my brother Martin exploded through the kitchen door after work. "Look at what my boss let me bring home for the weekend!" He shouted, "Come see this!" All my siblings and our parents gathered around the kitchen table. We stared at the dull pea-green colored typing machine with a rotating ball inside, and then Martin plugged in the new IMB Selectric typewriter. Comments ranged from "It's incredible," to my mother (who had worked as a secretary) huffing, "Great, now they can fire one typist because the secretary left behind will be able to do the work of two."

And so the information age began.

My mother's fear that technology would change the workplace and result in a change of society was and is real. Yet Dr. Brown's book is a guide to using the thousands and constantly-evolving TOOLS of technology without abandoning who we are as a human society. He provides resources within this book that give educators, business leaders, psychologists, "C" level executives and communication professionals the necessary tools to utilize technology intimately to foster and enhance human connections.

We have adapted and we have changed, but technology has not taken the deep need for human communication and social interaction away from us. In fact, Dr. Brown articulates how it has the potential to deepen ties and to equalize disparities in our world, from solar panels in African villages, to electronic commerce, to empowered communication between those who have shared values via Facebook.

The key, especially for older students and workers is to commit to adapt. We need to adapt not only to new technologies, but to the younger generations who already have different neural pathways that accommodate new technologies as part of their culture and lifestyle. Dr. Brown has provided one of the first guides for older readers to comfortably make that absolutely necessary adaptation. Know this, as you read the text you are utilizing the plasticity that is within your neurotransmitters. Adapting communication so that it is meaningful requires new skills, (not fears) that need to be learned and embodied by older workers in all societies.

Are We Too Hardwired to Change?

Evolutionary psychology as a discipline is now a well-accepted convergence of research and discoveries in genetics and neuropsychology. *Solutions for High-Touch Communications in a High-Tech World* acknowledges that modern human beings inhabit a thoroughly modern world of cyberspace and virtual realities, but psychologists and neuropsychologists will tell you that ALL of us have an ingrained mentality for adaptation and that people today still seek those deep and basic traits that made survival possible from the ice ages through the new millennia. We are, in other words, hardwired for adaption and change. Dr. Brown's publication is a concrete pathway that will lead thinkers and early adaptors to that change. His publication enables the successful combination of technology with emotional intimacy. It is possible and, yes, it is teachable.

Just as print media has radically changed to accommodate social media, Dr. Brown's text explores and teaches the newest concepts in social media theory so that they are understandable to the reader.

Social exchange theory is a way to look at patterns and the rules of conversation. It assumes that unstated rules of communication must be followed if the conversation is to be a success. This is a massive consideration when the communication comes from the Internet where the communicator is not visible and is shielded by their computer. Yet in Brown's book, he demonstrates that transparency is required, unless the sender of the message wants to be placed in the awful situation of presenting themselves as something or somebody they are not. One of the greatest gifts of the social media world is honest, genuine communication.

Relevancy is also required for successful social media communication. Since the receiver can CHOOSE to communicate with the sender, all communication must be relevant, and therefore two-way. This achieves very quickly another, deeper level of social and person-to-person intimacy. Dr. Brown articulates that successful social media requires communication specific to the topic. Tangents confuse the receiver and therefore are naturally avoided at all costs. So communication is again deeper, more meaningful, and more relevant.

Segmentation of communication makes communication easier. It is commonly accepted that blogs, Twitter, and Facebook are for personal use. It is typical to put information relevant to one's personal life on these sites. These connections are new and help connect families and friends and locate social "villages" with shared values. Key concepts of social exchange theory include reciprocity, fairness, and negotiated rules with information, approval, respect, power, group gain and personal satisfaction among the rewards in successful transactions.

Because of social media, relationships of any distance can become more intimate over time. Millions of friendships have formed from "tweet-ups." And many friendships evolve beyond the computer or Skype or Facetime because of these intimate exchanges.

Mara Woloshin
Woloshin Communications Inc., USA

Preface

I'm excited to welcome you to my project. This book is intended to provide vice presidents, managers, researchers, students and other academicians with a new way to approach the communication challenges by addressing the gaps and weakness of digital-only communications. But this book goes beyond academic and organizational utility to be a source that can help people every day in their personal communication activities. Whether you are communicating online with your family, or looking for a job, or continuing your education, or creating a new relationship, this book has something for you. *Solutions for High-Touch Communications in a High-Tech World* offers a new outlook on the necessity of face-to-face communication, which we will refer to as "FTF." This new outlook is presented without the intention to discard or minimize the need for and importance of online communication. This is intended as a cooperative effort. As we move through this learning experience about communicating face-to-face and online, I will treat your journey as a discussion. So "we" will experience everything together to share a different outlook on communicating in today's world.

There is an obvious gap between FTF and online communications. There are growing communication challenges posed by the evolving intergenerational use of social media and the need to reintroduce face-to-face communications in our world. Social media currently defines many cultural, social and work environment considerations for people and it is very powerful, but it inherently creates a cross-cultural and cross-generation communication problem. This is because the full range of the communication process is limited or absent in the online communication process. Digital communication has proven itself as a communications tactic with the ability to build trust among strangers, but there are much stronger bonds that are often required for successful face-to-face interactions. The fact that the bonds are missing creates the communication gap that we will examine in this book.

This work offers information to bridge the difficulty when individuals go from long-term or exclusively online communication to organizational and/or personal face-to-face communications. This is an important and specific challenge that very few experts are addressing. Forbes reports that by the end of 2016 some 63 million Americans will be working and communicating online, and large numbers will forego FTF communication. The High-Touch communications spelled out in these pages focus on fully employing the communication process to effectively move seamlessly between online and FTF interactions in the workplace and at home.

The basis of my beliefs about the communication gap started in May 2011 as I finished my doctorate in public administration and urban policy from Old Dominion University in Norfolk, Virginia. My dissertation and the connected research form the basis for developing this brand new approach for emphasizing face-to-face communication, creating social bonding, developing social capital, and taking advantage of the full communication cycle.

You are about to embark on a journey to learn how people and groups can determine a communication "end state" and then devise a strategic approach to modify communications to meet the end state goals. We will explore how the reader can utilize expectation, value and return (EVR) in social and interpersonal networking to successfully fuel the "end state" approach. We will analyze ways to leverage my findings on effective Social Networking and Individual Performance (SNIP) to improve communication and build effective groups characterized by long-term relationships. Figure 1 illustrates the characteristics involved in high touch and high teach activities.

New Age Communications Overview

When people talk about social media, they refer to the different tools or platforms one uses to communicate socially. Social networking activities use those tools or platforms, social media, to share experiences, get advice, make decisions, develop trust, and reduce risk. So social networking is about our actions and social media is about the tools we use to energize those actions. In the truest sense of social networking, people are motivated to contribute valuable information to a person or a group. All communicators should ensure two-way interaction whenever possible.

Another communication consideration is information appropriateness, both contextual and cultural, which is about ensuring that topics of conversation are relevant for the receiver or group. Relevance and appropriateness are based on demographic values of the community, as well as cultural considerations.

Figure 1.

In other words, people and groups should act and communicate in ethical and transparent ways.

Section 1, "Social Communication," defines the communication gap between FTF and online or digital interactions and the characteristics of each method. It talks about bridging the communication gap and provides a short history of computer mediated communications (CMC). Motivation, culture, and multi-ethnicity are also covered in the first section. Current research in online interactions is covered in this section, followed by details on SNIP.

Section 2 asks "How do we connect?" This section tries to answer the question with in-depth comparisons of FTF and online communication. This is followed by information on interpersonal and group communication.

Section 3 introduces New Age Communications as well as engaged interaction, which is about senders who are completely tuned into communication and who are willing to make the necessary changes to ensure communication success. New views on FTF and online interactions are shared next, along with suggestions for improvement and resources to assist in the each method. Section 3 ends with an explanation of the hybrid approach to interpersonal communication that was intended for this work.

Let's look at some specifics of what we'll cover using Table 1. Everything in these pages is based on the premise of communication that involves a sender who takes an idea, encodes it, and sends it to a receiver who decodes it. We will also focus on feedback in great depth. As communicators, whether face-to-face or online, we need to do more than just send. We need to interact and deliver value that convinces the receiver to take action. Online communication, blogging for instance, refers to this as a call to action. We will also examine social cues, non-verbal cues, and shared meaning.

Call to action: This should motivate the receiver to exhibit a behavior and specifies what action should be taken to elicit that behavior. The call to action might be something like "Click here to buy now," "Enter your email to join

Table 1. Analyzing communication considerations

IMPROVING COMMUNICATIONS
Value Creation • Expectation, Value and Return (EVR): focus is on getting useful help in a two-way communication arrangement, achieving a valued information exchange, and getting and giving recognition that matters to others. • Social Networking and Individual Performance (SNIP): provides information, techniques and examples that address real or perceived benefits in order to assist in the social networking journey. • Two-way Interaction: sender remains engaged to ensure message is decoded and the receiver provides feedback. • Information Appropriateness: share information that matters to all involved.
Skills 1. Social Communication: high-touch and high-teach approaches to interaction. 2. Communication Theory: determine how to connect whether FTF or online. 3. Engaged Interaction: senders must tune in and stay tuned in to ensure success.
Resources • Sender-receiver relationships: interaction, delivery of value, call to action, and mutual agreement. • 4Rs of New Age Communication. • Key Theory: Emotional intelligence, Sensemaking, Communication Accommodation Theory (CAT).
Engagement 1. Interactionism leads to Engaged Interaction a. Interactionism is when people act based on social and contextual cues they derive from exchanges with others. b. Engaged Interaction is when sender and receiver fully commit to listen, hear, and understand and commit to staying engaged until success is achieved.
FTF & Online 1. A deeper understanding of face-to-face and online communication; 2. A focus on the communication process gap between the two methods; and 3. Knowledge leading to resources to "bridge the gap" and create the best messaging interactions possible regardless of the method.
High-Touch & High-Teach • A more robust understanding of communication and a focus on the gap in communication are ***high-touch*** issues relating to getting people to interact. • Resources to bridge the gap are ***high-teach*** considerations leading to a plan for changing and improving the way we communicate.

today," or "Use our secure form to order." Done well, this gives the receiver a sense of urgency that matches the intent and goals of the sender. Sense of urgency comes from incentives, like offering a discount, gift, or free trial in return for taking the specified action. Senders, such as advertisers, are also trying to create a sense of urgency when they inform the receiver that the items are unique or that they are only available for a limited time in a limited quantity. The best and most popular approach for the call to action seems to be giving a deadline and warning the receiver that they will pay more or lose availability completely if they don't take the "recommended" action by a certain time, or before the limit on numbers of respondents is reached. The best calls to action are clear, direct, and decisive: write to Congress today, start that master's degree program right away, or improve your communication skills by participating in face-to-face communication at every opportunity.

Social cues: The generation of ideas should lead to improved communication that shares information with value to all participants and that conquers communication barriers. Social cues allow improved interpersonal communication and build social networks that improve the way we interact. We must conduct communication activities that work for a guarantee that the receiver got the message in the way it was intended and that they appreciate the value we intended to provide. The mutual agreement between the communicating parties is the foundation we need to create so that our interactions are continuous and cyclic. We must strive to get all communication participants to accept the responsibility to be genuine and accurate, and we must allow engagement through the use of visual, verbal and non-verbal cues. We must work hard in FTF communication and harder in online conversations to make the connection.

Non-verbal cues: The connection we want to make has many moving parts. Our premise in this book focuses on non-verbal cues that are available, but that are not used in online communication. This is a powerful tool for communicators and the reason this journey advocates a greater focus on face-to-face activities. It may be artificial, but online communicators would be well served to find ways to use non-verbal cues because they are a fundamental part of human development. Non-verbal communication occurs naturally as we grow, appearing during infancy and early childhood, according to many experts. It allows us to "read" people and get a sense of their approval, disapproval, discomfort, joy, surprise, etc. Non-verbal cues allow us to conduct full-range communication because they give us more data with which to get feedback and send additional or adjusted messages.

Shared meaning: It's important to get your message across the first time you send it. Sometimes repetition helps, but only if there is shared meaning that leads to a continued interest by the receiver. Great communications are empathetic, authentic, and transparent. Senders should be genuine in their interest in the message to the receiver. They should stay connected to the receiver to ensure both parties are getting value from the interaction. Transparency allows the receiver to see and understand all actions of the sender. This builds strong engagement bonds that lead to trust, respect, and support between the communicating parties. Senders should be in touch with their own thoughts and emotions to promote an ease of conversation that will allow others to truly share in the interaction. Effective communicators have a way of understanding beyond the conversation. The ability comes from being self-aware and using the whole range of verbal and non-verbal messages in the communication situation. Effective communicators can decipher confusion in the interaction. They can figure out what is unclear in terms of statement, witness, or discussion. The transmission by body language and other non-verbal cues requires an effective sender in order to be decipherable.

A great focus on communication interaction does not completely ensure success. When there are difficulties, senders can use or reuse the 4Rs of New Age Communications to improve or reshape the difficulty: Restart, Revise, Reorient or Reschedule. These are approaches that can be used to energize the communication while it is in progress. *Restart* the conversation by going back to the beginning of the interaction and getting a fresh start, or by rewinding a bit and focusing your efforts. Restarting gives both parties an "escape" so that they can change the communication in a positive way. Restarting leaves the past behind as much as possible, clears the air, and charts a new course. *Revise* the conversation by changing the focus, the tone, or the language you're using to get better understanding and seek agreement on your message. If the sender is not coming to a "meeting of the minds" with the receiver, a great approach is to change the focus of the communication. *Reorient* the conversation by presenting the same message in a new or different way to improve reception and enhance decoding. Use your imagination, be creative, and change the conversation to ensure the message can be decoded by the receiver. *Reschedule* by trying the conversation at another time that is mutually beneficial, or by trying it with another receiver. You might even try the conversation in another setting. Pick an approach in difficult situations and then pay attention to the results, avoid distractions, and ensure you make the connection. Help the communication by adding value, understanding the concerns of all parties involved, and meeting the intended outcome of the interaction.

To enhance the learning adventure and arrive at true communication interaction, this book provides a full range of information on communication, from history to theory to interactivity comparisons. A short history of online communication is provided because technology now dominates much of our interactions. Experts tell us that many people rate FTF communication fourth behind phone, email, and instant messaging. Millions of people have signed up for the portable communication offered with the smartphone and tablets. The fast development of high-speed, large-bandwidth mobile networks means that half of the world's population now has a mobile subscription. That has increased from 1 in 5 just five years ago and some experts project the global penetration rate to increase to 60% by 2020. This work provides valuable information on improving communication through motivation, multicultural awareness, and attention to improving diversity leading to social cohesion.

We will provide relevant research and applicable theories to help with the concept of bridging the communication gap. Theory is important to help us explain behavior or performance in an interest area, and then facilitate predictions about that behavior. References to research results and connections to theoretical understanding helps communicators devise ways to bring about change after we explain what we see in the literature. The theory we use allows us to identify the problem and then compose a plan to change the situation. Theory also helps us justify our stated position on an issue or problem.

Many theories are covered in this work, but there are three that really provide a foundation for comparing and contrasting face-to-face and online communication: emotional intelligence, sensemaking, and communication accommodation theories (CAT). Trust will be examined in several parts of this work and that's important because it helps us with social context cues. Using trust in communicating requires that we focus on its use to deal with uncertainty and to address acceptance of vulnerability.

Next, we explain full-range communications and how to achieve their use. We know that communication is an interactive process to exchange messages and deal with noise in the channel. We compare and contrast FTF and online communications and apply theory to what we discover. This delivers the skills necessary to move from FTF to online, or to combine the two methods for the best effect. The goal is to always engage the human element in FTF, and to find a comfort level so online communicators will leave the computer and employ in-person interactions fueled by verbal and non-verbal cues.

Six theories that support interpersonal communication provide relevance for our approach to interaction. We define interpersonal communication as a direct, in-person exchange between participants in an interdependent relationship. We address communication necessities: feedback, resolution of conflict to save face, correct the record, determine winners or losers, or to set terms for future interactions. Our educational journey also analyzes how the theoretical foundations can be applied in either communication method.

Communication requires commitment in terms of dedicating time and attention to the interaction and managing the message, telling the truth, seeking shared value, and staying engaged to seek feedback. Committed communicators will do everything in their power to send the right message with the right delivery. Face-to-face communicators have the power to influence the conversation and motivate a person or group in order to achieve a goal or goals *DURING* the communication. This is not always possible in online communication.

As we reach the end of our educational journey, we present the theory of interactionism to demonstrate a hybrid approach that is neither FTF nor online in nature. We present a blended approach that energizes the best of both methods, giving communicators tools and resources they have never had before. This new approach to communication should allow senders and receivers to remove the barriers in both methods to achieve the greatest possible use of Engaged Interaction. It should focus all of the resources and tools provided to get the message across in the best way. It should help us build relationships and strong social bonds that guarantee full-range communication activities. Then we can listen, hear, and understand each other whether we are in the same room or worlds apart.

Our journey will allow readers and communicators to evaluate, dissect, define, and understand the communication gap that exists between FTF and online communication. We will examine theory, dispel myths, address barriers, introduce new concepts, and turn each method upside down. We will highlight the benefits and dangers, the pros and cons, and the ups and downs of each method. The hybrid approach presented in Chapter 11 is an excellent way for communicators to have success regardless of the method.

Solutions for High-Touch Communications in a High-Tech World is a communication journey that will give the reader the three valuable communication abilities listed below. A more robust understanding of communication and a focus on the gap in communication are high-touch issues relating to getting people to interact. Resources to bridge the gap are high-teach considerations leading to a plan for changing and improving the way we communicate. In the end, we offer a hybrid approach to communicating that is the culmination of the journey.

1. A deeper understanding of face-to-face and online communication
2. A focus on the gap in the communication process between the two methods
3. Knowledge leading to resources to "bridge the gap" and create the best messaging interactions possible regardless of the method

The hybrid approach is enabled by using the learning from all 3 sections of the book. Section 3 is called "New Age Communications" and it is intended to help build a foundation that businesses, educators, students, administrators, and other individuals can use to craft a plan for communicating. The plan would then bring all of the research and knowledge from Sections 1 and 2 together to help people communicate in a deliberate, efficient manner.

Great communication is about connecting with other people and creating and nurturing relationships. The information tour on which you are about to embark will arm you with plenty of information, research, resources, and skills. In reading this book, you will learn to communicate effectively across the "bridge" from the digital environment to FTF interaction. This is important because social networks can adversely affect communication due to decreased in-person conversations, superficial communication experiences, and the difficulties that digital communications cause in terms of degraded grammar and spelling and other language complications. These are important skills that are not always written about or taught in personal and organizational settings.

Finally, I'm very excited to have Mara Woloshin, APR, Fellow PRSA, write the foreword. A long-time member of the Public Relations Society of America, the largest PR organization in the world, Mara is the perfect choice to assess the value of this work. In addition to more than 20 years working in all aspects of health care marketing, she has spent 12 years specializing in aging medical relationship marketing. Mara has unique qualifications and is the consummate professional whose experience and long record of stellar accomplishment make her the perfect choice to examine and discuss this book.

I hope you enjoy this educational journey. Communicate, interact, and create great relationships. Good luck!

Acknowledgment

I thank my family for being there and putting up with me through an exciting journey that was difficult, frustrating, and exhilarating all at the same time. I couldn't have done this without you.

I want to thank, again, my good friend Mara Woloshin for helping with the book from proposal to writing the foreword. Mara is a wise colleague and friend who came through every time I needed help.

Thanks to Nick Webb for creating the outstanding graphics and color schemes for this book, our second time collaborating.

Introduction

Welcome to an adventure that will combine high-touch and high-teach approaches. High touch is about getting in contact with people, whether face-to-face or online, and communicating in the best possible, two-way interactive manner (Figure 1). This book contains a communication journey that will give the reader three valuable communication abilities:

1. A deeper understanding of face-to-face and online communication,
2. A focus on the gap in the communication process between the two methods,
3. And knowledge leading to resources to "bridge the gap" and create the best messaging interactions possible regardless of the method.

Figure 1.

When you speak to me, see the real me, the whole me, and the interactive me. It is only then that we can truly communicate. -Dr. Michael A. Brown Sr.

Sections 1 and 2 are the high touch part of this work, covering research and knowledge about communicating. Section 3 brings in high teach in that it addresses how to navigate the communication gap between the two methods and provides help to improve communication activities.

The information contained in this work is intended to reinforce education and training in communication, social media, interpersonal issues, team building, and trust for organizations and academia. At the outset of the creation of this work, the value was intended for public administration environments and academia; however, the learning experiences can be used in any number of industries and settings. This book is a unique approach to the communication process, applying that emphasis to social networking and social media activities in a holistic process that compares face-to-face (FTF) and digital interactions. While many books on communication address either traditional communication or mastering online communications, this work addresses a communication gap that exists between the two and then links them together. Readers will learn to communicate effectively across the "bridge" from FTF interaction to the digital environment. This is important because social networks can adversely affect communication due to decreased FTF conversations, superficial communication experiences, and difficulties that digital communications cause in terms of degraded grammar and spelling and other language complications. These are important skills that are not always written about or taught in personal and organizational settings.

This work will be presented as a conversation. We will have "discussions" about communicating. The educational journey will be successful if we embark on it with open minds.

Section 1
Social Communication

Chapter 1
Bridging the Communication Gap

OVERVIEW

This chapter will address the beginning considerations in our FTF versus online analysis, examining the gap that exists between the two based on the process. There is a challenge involved when moving from strictly online interactions to FTF, and that challenge has to do with use of the full range of the communication process from sender to receiver to feedback.

Before we go further, please understand that we will discuss social media at various times in this book. We will use online communication as the umbrella term to encompass all of the technology-based interactions we want to cover. Many experts say that social media is a part of online communication, which is, in turn, a part of electronic communication. Since online communication is now a widely accepted term, we will use it to refer to social media and other digital conversations. While we recognize that there are intermediate forms of communication between FTF and online (letters, faxes, telegrams, etc.), we do not address those.

This chapter will cover the importance of the feedback cycle, shared understanding between sender and receiver, and the use of social cues to improve or enhance the communication. There is a new set of communication barriers, formed in large part by the emergence of the digital conversation. Finally, the chapter offers an understanding of the characteristics of social networks and interpersonal communication and how they interrelate in today's interactions. The feedback cycle is crucial in its communication importance because we know that messages are rarely fully "one-way" activities. The sender needs

DOI: 10.4018/978-1-5225-1897-6.ch001

clarification that the receiver or receivers decode the message in the way it was intended or that they received the value that the sender intended to provide in the communication. A mutual agreement between the communicating parties is vital to ensure effective interactions that are true to the original message and its meaning, and that set the state for the continuing, cyclic conversation that should follow the original engagement. A truly effective communication features all parties accepting responsibility to be genuine, accurate, and to allow every opportunity for human contact or interaction to be part of the engagement. It's also important to consider social cues, which are very powerful communication tools. Non-verbal communication normally occurs automatically, appearing during infancy and early childhood according to many experts. We "read" social cues to allow us to get a sense of people and improve the communication experience. Finally, social networks often lack the full range of the communication process. This is because they can adversely affect the interaction by decreasing FTF activity, imposing superficial communication experiences, and contributing to grammatical and spelling erosion of the language. So, our challenge becomes whether we truly get feedback and important communication value from the interaction.

Communication Challenges and "The Gap"

Now we begin our examination of the communication challenges posed by the ever-growing use of social media, almost exclusively for some. While social media and digital conversations are powerful, they do not go far enough in taking advantage of the full range of the communication process. Electronic communication is evidence of the ability to build trust among strangers, but there are much stronger bonds available that can be gained by face-to-face interactions. The problem is that many people need and want the ability to build a true team out of people who have never met. In that first meeting, there is a communication gap.

There is an obvious gap between FTF and online communications. There are growing communication challenges posed by the growing communication challenges posed by the evolving intergenerational use of social media and the need to reintroduce face-to-face communications in our world. Social media currently defines many cultural, social, and work environment considerations for people and it is very powerful, but it inherently creates a cross-cultural and cross-generation communication problem. This is because the full range of the communication process is limited or absent in the online communication process. Digital communication has proven itself as a communications tactic with the ability to build trust among strangers, but there are much stronger bonds that are often required for successful face-to-face interactions. The fact that the bonds are missing creates the communication gap that we'll examine in this book.

As our cyber workplace grows, many managers are not able to recognize the need of individuals and staff members to have new tools to build a truly effective team. Virtual teams can be hindered and fail to achieve maximum efficiency if they have never met. They can be successful, but this book maintains that the interaction is more complete and more powerful when it is face-to-face. So, that first face-to-face team meeting (if it ever happens) creates a communication gap because not all members are equipped to employ the full communication process involving sender, message, receiver, and feedback. If not successfully addressed, there is a likelihood that the team will not function as efficiently as possible. The danger is that there may never be a meeting of the minds, which is crucial to team and culture building.

Addressing the gap is not a battle cry for one method or the other; it is an observation that FTF is beneficial and inevitable and requires skill. Further, it is a statement that online communication would be improved by employing some characteristics of FTF, like feedback, to bridge the gap between the two methods. FTF is a "high-touch" method that works to apply the full communication process. This book is a "high-teach" approach to learning how to bridge the gap between the two methods. We will cover positives and negatives with both methods, emphasizing the things that are difficult when transitioning from the online method to FTF. Then, readers will be armed with the tools they need to conquer the communication gap.

Because digital communications are so prevalent and powerful now, you might think the communication gap isn't really a problem. But leaders, managers, and supervisors would tell you differently. For instance, studies on Gen Y and Millennials demonstrate that these groups will comprise more than 50% of the workforce by 2020, and many experts have found that these individuals prefer to use instant messaging or other social media rather than stop by an office for a talk. That preference makes it a real challenge to build and nurture a team and corporate culture.

For example, a 2012 study by United Kingdom researchers analyzed how the presence of mobile communication technology influences the quality of face-to-face conversations (Przybylski & Weinstein, 2013). Their findings suggest that developing the interpersonal closeness and trust that is required in face-to-face communications is made difficult by the mere presence of mobile phones. And the results suggested that the difficulty is even more pronounced when the discussion is of a personally meaningful nature.

In 2010, a national study by the Kaiser Family Foundation examined a sample of more than 2,000 eight- to eighteen-year olds. The study discovered a rise in the average total time that they reported experiencing media (Rideout, Foehr et al. 2010; Pea, Nass et al. 2012). The study found that the average rose to 10 hours and 45 minutes per day in 2009, up from 8 hours and 33 minutes in 1999, a 44% increase over a decade. The proportion of media time spent using more than one medium concurrently increased from 16% in 1999 to 26% in 2004 to 29% in 2009 (Rideout, Foehr et al. 2010; Pea, Nass et al. 2012). Obviously, the numbers must be even higher now.

Social media allows us to communicate on very serious topics like contracts, purchases, and love without the benefit of verbal and/or visual cues. Even when using video, it is evident the face-to-face interaction can be much more effective when one can use all five senses. These are important communication considerations when we think about the sender, message, and receiver interchange. They are also vital considerations when changing a communication experience from totally online to face-to-face.

As we continue to examine FTF vs online communication, we have a problem. There are a number of studies relating to learning and instruction that is technology-based versus in-person, such as the work some researchers did in comparing course instruction (Stacey & Wiesenberg, 2007; Jaggars, 2014). However, this learning adventure is not about the kind of pure communication that we are discussing and that is found in many of our references on trust, comparing the two methods, social skills, interpersonal communication, and mobile technology (Ma, 2007; Baek, Wojcieszak et al. 2012; Brown, 2013; Przybylski & Weinstein, 2013; Gheorghiţă & Pădureţu, 2014; Carter & Fuller, 2015). This book addresses all communication, not just that which occurs in an educational environment. This is a topic which many researchers, including this author, believe is not being addressed, studied, evaluated, or dissected appropriately. This book is a first step.

So, let's examine how to get ready to experience the whole communication process in social media. Along the way, we'll examine how to help prepare your online communications to complement the face-to-face interactions that are inevitable when building and improving trust. Some of the tips include asking very pointed questions, repeating answers, and doing follow up to ensure there is a shared meaning. Most importantly, you will experience how expectation, value, and return (EVR) can improve the communication task at hand.

Expectation, value and return, or EVR, is about using three keys to develop a winning social media plan that develops itself and encourages others to join your network. This is how you demonstrate the WIIFM – what's in it for me – to your intended audience, whether they are one person or a group. In the true sense of social media interaction, people are motivated to contribute valuable information to a person or a group because of an EXPECTATION of VALUE in RETURN. We're talking about three things. First is the ability to get useful help in a two-way arrangement. Second is the opportunity for a valued information exchange. The third benefit to EVR is getting recognition that matters and giving it to others Of course there's more to social media participation than just understanding EVR. But EVR gives you the tools to effectively connect with others.

Education is another place where we see the importance of having at least some face-to-face interaction in the online classroom. In fact, organizations of all types interested in transforming interpersonal communication and team building are turning to hybrid formats. These leaders are now adding blended interactions that combine online and face-to-face sessions for the best effect.

A 2012 study by the non-profit ITHAKA evaluated the quality of education that holds the promise of broadening access to higher education to more individuals, while also lowering costs for students. "Our intention was to provide a rigorous side-by-side comparison of specific learning outcomes for students in this hybrid version of the statistics course and comparable students in a traditionally-taught version of the same course."

Hybrid-format students did perform slightly better than traditional-format students on three outcomes and, while none of these differences passes traditional tests of statistical significance, there is merit in the notion of educational flexibility.

In 2014, Cowan and Menchaca evaluated value creation in a hybrid online graduate education program. The study demonstrated value in the hybrid environment evidenced by the 10-year success of a master's program in educational technology. In survey responses, participants specifically mentioned the importance of:

1. Problem-based learning,
2. Connections to real-world problems,
3. A flexible learning environment relying on many newer technologies, and
4. The critical importance of a face-to-face component in establishing community.

Let's take a deeper exploration of the communication gap and its challenges.

Feedback Cycle, Understanding and Social Cues

The feedback cycle is very important in the communication process because messages are rarely sent in a "one-way" fashion. The sender needs to know that the other party or parties interpreted the message correctly, or that they received the intended value from the message. In management issues, feedback is necessary because the leader needs to know how subordinates respond to directives and plans. There is a need to understand how work is progressing and to get a sense of how employees are invested, or not invested, in the work environment.

The need for a common understanding is the cornerstone of communicating. It is critical to measure effectiveness through a mutual agreement between the parties about the message and its meaning. Effectiveness is achieved when there is an indication that the sender understands the message in the manner intended by the sender. This should be the goal of any communication, but people sometimes assume or take for granted that this agreement is reached.

Common understanding, or the lack of it, is a weakness of online communication because there are limitations or absences of social cues and shared experiences. These are important to the meaning of the message and the ability to decode and provide feedback. These issues can be minimized, but not necessarily overcome, through use of the full-scale communication opportunity of face-to-face interaction. For instance, to achieve deeper meaning in communications, one of the requirements is a level of trust between the parties. It is easier to build trust and in turn social capital in FTF interactions because there are more tools to help the process in this form than in online communication.

When the communication features a human, "hands-on" interchange, we can use verbal and visual cues and the five senses to achieve our goals. These tools are not all available in digital conversations. One example of the many challenges is when prospective members of a team must meet face-to-face first, then use electronic communication to assist in the process of team-building (Sproull & Kiesler, 1991, 2000; Olson & Olson, 2000).

Common Understanding and Mutual Agreement

Communication feedback should be accomplished through a continuous loop between the parties involved. It should start when the message is received and subsequently decoded. After that, it should reengage the communicative process, generating a newly-encoded message as the receiver becomes the sender. This cycle is vital to providing information that can enhance, clarify, or restate the previous message. In this way, feedback fosters common understanding and mutual agreement.

The best communications feature sender and receiver understanding the message in the same way at the same time. This approach allows a deeper understanding of the communication and increased ability to share with others. Great communication also requires mutual agreement on the topic, course of action, or necessary changes to achieve a goal. What all of this means is that good communication requires collaboration between sender and receiver. The process finds success in digital and in-person interacts all the time. However, when there are difficulties with understanding or agreement, in-person conversations have tools that may not be present in the digital environment.

Social cues are known to be very powerful in communication. Our non-verbal communication skills are a normally an automatic occurrence, coming into play during infancy and early childhood, according to many experts. Social cues are skills that help us get a sense of people, or "read" them. They include, for instance, using gestures, engaging in non-verbal interaction with another person, or reading facial expressions and "hearing" tone of voice. These skills are not directly taught and they develop primarily in an unconscious manner. These are face-to-face skills for the most part and, as we'll learn throughout this book, they can atrophy over time in the digital conversation because they are not normally used. More important for our communication endeavors is that these skills might not develop fully, or might be lost, in the digital environment.

Social cues help us respond to the receiver's tone of voice or facial expression. Social cues arm us with gestures we can use to get help, seek approval, and express agreement. Without these cues, or without the full use of them, the sender doesn't receive vital information, such as whether a situation is dangerous or safe. Without these cues, senders lack the ability to work through language issues. Without these cues, senders are limited in their ability to get and keep the attention of receivers. Without these cues, we struggle to participate in a successful interaction with another person, and we blind ourselves to the non-verbal world. These situations may get more serious when strictly online communicators must complete a face-to-face interaction.

Some might argue the importance or relevance of social cues in digital communications, but without these cues there are simply gaps in the interaction. For instance, without an emoticon or ALL CAPS, how do you decipher when the receiver is mean, happy, sad, scared, or agitated? Without postures, gestures, and movement, how can you tell when a person is uncomfortable or anxious to leave the communication? We also need to take into account tone of voice, which provides as much information as the sender's actual words, and sometimes even more. Tone of voice allows us to make a communication evaluation of several things such as volume, tone, and rate of speech. Also, eye contact provides very definite cues in terms of what the receiver is thinking. Do they stare straight at you? Do they look away? Do their eyes wander around the room? Are they rolling their eyes to show disconnection with the message?

The "traditional" way to refer to online communications is a reference to computer-mediated communication (CMC). Another term is computer-mediated relating (CMR), which emphasizes a larger area of interest that includes varied interpersonal dimensions of interactions (Cooper & Sportolari, 1997; Whitty & Gavin, 2001). CMR encompasses a broader spectrum of interactions, such as online friendships and romantic communications. Some would argue that online relationships, for instance, lack depth and meaning when compared to those conducted face-to-face because they don't take advantage of the full "bandwidth" that is available. The current discussion differs from the normal bandwidth argument, in that our position centers on the benefits that can be gained by using more bandwidth, especially when one takes an online conversation into a face-to-face setting.

There is an intimacy without emotional investment in online interaction that can lead to strong relationships. Authors Monica Whitty and Jeff Gavin write "This situation is demonstrated in the 'boom or bust' phenomenon, in which a rapid process of intimate self-disclosure leads budding relationships to become quite intense quite quickly" (Whitty & Gavin, 2001). Whitty and Gavin discuss a communication evolution that demonstrates an accelerated process of revelation that increases the chance of exhilaration at the outset, followed by a quick erotic episode. Unfortunately, they contend that this relationship cannot be sustained because it lacks the underlying trust and true knowledge between the participants that is required to support long-term interaction.

The Chapter 2 examination of theory will include social presence theory. In short, this theory evaluates "the degree to which a person is perceived as a 'real person' in mediated communication" (Gunawardena, 1995). The theory focuses on intimacy in communication and psychological distance. Using social presence theory, the argument is not that online interactions completely lack these characteristics; it is that they are not as strong or fully formed as in face-to-face interactions. There is an obvious difference in the amount and quality of social information between communication participants in these two forms. In terms of CMR, lower levels of social presence make for more impersonal conversations. The lack of contextual cues can also hinder online communications.

The absence of social content cues can lead to more uninhibited behavior, such as verbal aggression, blunt disclosure, and nonconforming behavior. This type of behavior, known as "flaming," has been observed across a range of online settings, including business, governmental, educational, and public networks. Such findings lend further support to the argument that online relationships are less inti- mate and more aggressive than face-to-face relationships. (Whitty & Gavin, 2001)

Some argue that certain benefits of online communication, such as trust and relative anonymity, would give it advantages over face-to-face interaction. For instance, social distance has been increasing since the telegraph and telephone were invented. Every day, as technology allows us to be anywhere and communicate online, social distance is strained. However, there are many rich and valuable communications that take place despite social distance considerations. In fact, there is an argument that internet conversations create very different first impressions of people than those in face-to-face, where physical attractiveness, gender, and age are immediately revealed or implied. This, online proponents argue, provides opportunities for more radical opportunities in relationship development than face-to-face does. Still, lasting relationships and complex communications lend themselves more to face-to-face interaction than online.

Trust in Communication

Trust is important in building relationships that are necessary for good communication, because trust allows the participants to develop dependence, satisfaction, and commitment. Trusting relationships in organizations involves an ongoing decision to give most people the benefit of the doubt, and it can be extended even to people one does not know from direct experience (Brown, 2011). There is a rich body of literature to help with the examination of trust. A group of guest editors addressed the widespread influence of trust in a special issue on trust in an organizational context (McEvily, Perrone et al. 2003). The discussion addresses the importance of trust, examining its current importance and finding that the current emphasis on trust was caused by changes in technology. At that point, the changes they examined reconfigured exchange and the coordination of work across distance and time.

The changes are still evident today in personal and organizational forums and in knowledge-intensive organizations. The group of editors explain that new ways of communicating "alter the patterns of interdependencies and the nature and extent of uncertainty" (McEvily, Perrone et al. 2003). These new ways make people rely more on the decisions and actions of others, creating both dependency and vulnerability. These factors are crucial to building and maintaining trust, and senders must pay attention to them. The authors continue to demonstrate that organizational science has made some important advances that promote understanding of the meaning of trust and how it relates to certain factors that characterize organizations and that are important for the current communication discovery. An increasing number of journal articles and special issues (Rousseau, Sitkin et al. 1998; Bachmann, Knights et al. 2001) and books (Gambetta, 1988; Kramer, 1996; Lane & Bachmann, 1998) are devoted to the topic of trust in and between organizations. The special issue published seven papers that represent a wide range of methodological approaches, a diverse set of theoretical disciplines, a variety of levels of analysis, and a blend of empirical models (McEvily, Perrone et al. 2003). Despite our primary focus on individuals, these views on trust are valuable in this context.

In the special issue mentioned above, two papers are important. Two researchers (Becerra & Gupta, 2003) probed the influence of organizational context on trust, emphasizing how the influence of social structure in an organization is contingent on communication frequency. They argue that frequency of communication is related to emphasis, in that as frequency increases, the emphasis shifts from the trustor's to the trustee's individual and contextual characteristics. The relationship among senior managers of a multinational organization is the basis for testing the hypotheses. Findings point to a view of trust production in organizations that consists of individual, dyadic, and contextual components.

A second paper argues that people tend to trust members of their own organizations more than they trust people from outside of the organization (Huff & Kelley, 2003). The authors present the notion that the effect of this trust is greater in collectivist societies than in individualistic societies. Trust is lower on average for people from collectivist societies, a prediction that runs counter to conventional wisdom (McEvily, Perrone et al. 2003). The authors generally find support for their hypotheses using data from a large sample (Huff & Kelley, 2003).

There are two important issues that are central to any examination of trust. The first involves trust as a means for dealing with uncertainty. The second focuses on trust and acceptance of vulnerability (Newell & Swan, 2000). Trust occurs in situations of risk and uncertainty, because any system must have trust as an input condition to stimulate supportive activities in situations with these two factors (Luhmann, 1988). This understanding suggests that trust is an attitudinal mechanism that allows individuals to subjectively assess whether or not to expose themselves to situations where there may not be an acceptable trade-off in terms of possible damage versus received advantage. The attitude develops when individuals have accepted vulnerability to others.

Trust is multi-dimensional and it helps values, attitudes, emotions, and moods interact (Newell & Swan, 2000). There are three reasons someone may be able to develop trust (Sako, 1992):

1. Because of a contractual agreement that binds the parties in the relationship;
2. Because of a belief in the competencies of those involved; and
3. Because of a belief in the goodwill of those involved.

Our communication needs require that we also consider other dimensions of trust, and there are a number of very important studies to help us understand it. It is helpful to examine trust in organizational settings, using rewards to increase and decrease trust, and examining the effects of third-party relationships on interpersonal trust (Dirks & Ferrin 2001; Ferrin &Dirks 2003; Ferrin, Dirks et al. 2006). A typology distinguishing between deterrence-based trust, knowledge-based trust, and identification-based trust was developed in 1992 (Shapiro, Sheppard et al.). A developmental focus was introduced in the late 1980s (Zucker, 1986), establishing three central mechanisms of trust production: process-based, characteristic-based, and institutional-based. Process-based focuses on reciprocal, recurring exchange. Characteristic-based is defined by social similarity. Institutional-based is determined by expectations embedded in societal norms and structures.

We can get a more personal view of trust than what has been discussed so far by focusing on intra-organizational trust, which is about individual performance in formal and informal organizations. Formal organizations would be your job or your political organization. Informal organizations could be your church, your softball team, or your volunteer group. Intra-organizational trust concerns interactions and activity within organizational relations, according to authors Chris Grey and Christina Garsten (2001). They argue that most literature gives more emphasis to inter-organizational trust, between-organization trust, but there are a number of researchers interested in what happens within organizations (Mayer, Davis et al. 1995; Li & Betts, 2003; van de Bunt, Wittek et al. 2005; Steinfield, DiMicco et al. 2009).

Later in this book, we'll discuss communication in group settings. There is a relevant 2010 study, especially in terms of online communications, concerning building trust in nonprofit networks (Lambright, Mischen et al. 2010). Factors influencing interpersonal trust in networks were examined in the study: propensity to trust, perceived trustworthiness of the trustee, the relationship between the trustee and trustor, and third-party relationships. The findings demonstrate that trust between a trustor and trustee positively influences expected future cooperation. The study can be related to strength of ties (bonding, bridging, or linking) in the network (Hatala & Fleming 2007; Ivancevich 2008), and identifies focusing on successful past cooperation, and frequency of interactions as two of the most important factors influencing interpersonal trust in networks (Lambright, Mischen et al. 2010).

Social learning theory is one of the considerations used in this book, and it relates to trust by maintaining that behavioral change in organizations can be the result of vicarious learning through modeling. Trust can be achieved if there is a successful learning experience. The nature of the observed model in vicarious learning can influence the chances that a sender or receiver will imitate the modeled behavior and that it will lead to success (Manz & Sims, 1981). People will normally seek out the model who possesses the greatest interpersonal attraction (Bandura 1977a). People who are perceived as being successful and who exert great influence are deemed more credible than others (Bandura, 1969, 1977a). Recent literature reviews on trust support the idea that modeling-based training programs will be more effective if the models presented have achieved high status and competence (Goldstein & Sorcher, 1974; Rosenbaum, 1978; Weiss, 1978). In one study, subordinates showed greater similarity in behavior to superiors who were viewed as being competent and successful, indicating great possibilities for building trust.

The importance of trust and its consequences for management was introduced through a network analysis (Gursakal, Oguzlar et al. 2009). The study sought to determine which employees trust other employees on a personal level. Based on the findings, the researchers argue that trust can create effective cooperation within enterprises, thereby having a positive influence on performance, growth, and survival.

A New Set of Communication Barriers

Every relevant metric shows that we are interacting at breakneck speed and frequency through social media. But are we really communicating? With 93% of our communication context stripped away by digital communications that may or may not have feedback, we are now attempting to forge relationships and make decisions based on phrases, abbreviations, snippets, or emoticons, which may or may not be accurate representations of the truth.

In an ironic twist, social media has the potential to make us less social. It is a surrogate for the real thing. For it to be a truly effective communication vehicle, all parties bear a responsibility to be genuine, accurate, and not allow it to replace human contact altogether.

It is common knowledge that electronic communication has surpassed face-to-face methods. The new age of business communications and the lack of comfort with traditional interpersonal communication contribute to the rise of digital communication. Some experts attribute this move to the influence of Gen Y and Millennials, and many believe that these generations will comprise more than half of the workforce by 2020. These groups tend to favor the instant kind of messaging afforded by social media and digital applications over sitting down in front of another person and interacting. This reality causes a gap in communication ability and actuality that characterizes communications between Boomers and their younger employees and employers.

One barrier that communicators must face is the need to create cultural compatibility, or find common ground. Senders can foster long-standing, mutually beneficial relationships by achieving shared values in their communication. This enhances shared understanding. Another barrier, one that exists in FTF and online interactions, is resistance to participate based on fear, disbelief, and underestimation of the benefits of communication, lack of a sense of return on investment of time and effort, misunderstanding and misperception. In FTF communication, use of verbal and visual cues allows the sender to get a sense of the resistance barrier and work to determine which of the reasons stated is at work. This activity is much harder, or impossible, in online communication because the receiver has to reveal the issue. Certainly, if the receiver doesn't respond, there is an issue, but the reason for the situation is hard to determine.

Other barriers are not new. Senders must deal effectively with the pace of communication to ensure that information exchange is timely and holds the receiver's interest. Both types of communications require attention to physical barriers. In face-to-face, the setting must allow a level of comfort to enter and remain in the interaction. Online interactions require that the physical surroundings of the sender are conducive to getting the message across. All communicators deal with attitudes and language in their own way. Lack of common experience and information overload are other barriers at play in any kind of communication. There are other barriers to communication, but the ones mentioned here are most relevant for our examination.

Communication should be an agile, flexible activity that keeps the sender and receiver in tune to the goals of the interaction. Sensemaking theory tells us to pay attention to the creation of meaning and to the perceived role each member believes they accept, and this is central to addressing barriers to communication. The communicating parties need to work through each step in the process, using cues to explain what is happening and to adjust to changes or difficulties in the activity. Communicators also use cues to create stories that match the encoded message and allow action and learning to take place.

Communication barriers are also addressed through the emotional intelligence of the parties involved, with participants working to recognize, respect, and value honesty, energy, feedback, and intuition in the communication. Additional emotional intelligence features that are relevant to communication are emotional fitness (authenticity, resilience, and renewal of message and idea), emotional depth (assessing character of the parties involved and seeking commitment, drive, initiative, and/or accountability), and emotional alchemy (using all resources to transform lesser ideas into greater ones through the communication activity). These considerations are important to addressing communication barriers, but so is common sense. For instance, make sure the message has value and pay careful attention to finding the right key words, language, and tone to get your message across. Transmit the message using the most appropriate communication vehicle for the communication participant, and help the receiver accept and decode the message. Stay attached to the communication to ensure that the receiver has the opportunity to provide feedback, which signals that two-way communication has begun and can continue.

The key consideration in terms of barriers to communication is that the sender must recognize them and address them when and if they interfere with the interaction. Barriers are often the key to driving receivers away from the communication, causing them to either not fully understand the message or not stay long enough to provide the valuable feedback that keeps communication alive as a cyclical process. Effective communications feature shared understanding between the participants and it can't be taken for granted. Senders must pay attention to any signs and signals to determine whether the message reaches its target. The communication goal is simple: encode, send, decode, and provide feedback. Senders must accept responsibility for this whole process, and for continued communications. FTF interactions give the sender a chance to look the receiver in the eye and make adjustments in real time. Online communicators must look much deeper, but the signs and signals are there. The premise of this book is that in face-to-face communications, the sender has many tools with which to change the communication in progress by using verbal and visual cues. That same capability doesn't fully exist in online communication. What is more telling about online communications is that they happen so fast that the sender often has no time to even consider making adjustments. It's on to the next communication. This is the challenge of barriers to communication.

Social Networks and Interpersonal Communication

In addition to barriers to the interaction, sometimes called noise in the channel, social networks may adversely affect communication by decreasing face to face interaction, by imposing superficial communications experiences, and through grammatical and spelling erosion of the language. Therefore, current studies seek to capture data about the spread of social networks, their use and impact on interpersonal communication. More specifically, they look for the answer to the question: what is the nature of interpersonal communication that is found on social networking sites: personal, emotional, private or shared, informal, and public (Gheorghiţă & Pădureţu, 2014)?

These are important considerations in communication. Readers will learn to communicate effectively across the "bridge" from the digital environment to face-to-face interaction. This is important because social networks can adversely affect communication due to decreased face-to-face conversations, superficial communication experiences, and by the difficulties that digital communications cause in terms of degraded grammar and spelling and other language complications. These are important skills that are not often written about or taught in organizational and personal settings.

It is crucial that we explore communication and misunderstandings at all levels of an organization, for instance. This approach to bridging the communication gap is focused on demonstration and theory-driven approach with relevant skills to help everyone understand communication processes in business and personal life. We address how to get the most out of communications that involve cultural, racial, and gender differences whether communicating in electronically mediated messages or face-to-face environments.

So the next time you communicate online, ask yourself if you're getting feedback and true value from the interaction. You just might be missing a critical part of a message you need! Find the value in your communication by understanding the real differences and similarities in the two methods, and how to deal with them.

The author of this work is a lifetime communicator, so there will be no attempt herein to make readers choose FTF over online activity. There is a time and a place for each and each has its plusses and minuses. The current focus is to manage the bridge or the gap when going from one to the other. The primary consideration is moving from online to FTF, because those skills have eroded over time. In fact, there are many people who have grown up in the digital age who have never mastered or used extensively the full range of the communication process. There are millennials and other groups who simply don't do face-to-face interactions with any regularity. These senders do not regularly use verbal and visual cues because they are unavailable, unfamiliar, or even uncomfortable. There can still be communication success in both processes, but the resources provided herein seek to help in achieving good results.

The most obvious benefits to digital conversations are the pure speed at which they can occur and the lack of limitations on the times and places at which communication can occur. Senders and receivers have complete freedom in the timing of their response. Some experts say that online communicators benefit from being able to think about what they say before sending the message. We can edit and revise the communication before sending to make sure we get the right message out there. Digital communication allows participants to enter and leave the conversation at any time, often without the other party even knowing that we are gone. Online participants, good or bad, can interact privately or can hide or mask their identity. This allows these senders and receivers to avoid phone calls or other human interactions that may or may not be convenient, necessary, or desired. Participants can relay or forward messages verbatim. Despite online limitations for developing verbal, in-person communications, digital communicators can engage in that area at a later date (and this book can help). Online proponents believe that their communications make it easy to communicate accurately (say what we mean), to be genuine (mean what we say), and to filter out unnecessary details or negative messages (determine whether to say everything we think based on the situation). And, of course, digital interactions allow us to navigate great geographic distances easily without the limitations of time.

What are the challenges in the online environment? First, there is the impersonal nature of some communications, which can be caused in part by the fact that we don't know who is listening or if anyone is listening. While it is true that online communicators can check what they say before sending, the reverse can be true. The sheer speed and convenience can betray the sender and the message can be sent before the thought process is complete. The digital environment doesn't allow us to determine the receiver's tone or demeanor. Online conversations are not recommended for bad news, emotional messages, apologies, or sharing affections because the sender cannot engage to make sure the receiver understands the seriousness or gravity of the message and is relatively safe and comfortable with the information. Many people use online communication for these situations to achieve anonymity, reduce stress, and alleviate accountability. This communication examination does not recommend that path.

Experts used to say that 90% of communication is non-verbal, but digital communication makes that statistic unrealistic. When online, senders can't determine facial expressions, eye contact, tone, pitch, speed, or body language. Emoticons are supposed to help in this area, but their applicability is limited. What if you have a feeling that doesn't match any of the emoticons? Other ways to help include video messaging or skyping, but they still don't have the full capabilities available in FTF interactions. Also, it is difficult or impossible to convey compassion in the online communication; messages may come across as just words without real emotion. Finally, the human condition is missing in a lot of ways when communicating online, but that is somewhat balanced by the method's ability to promote faster connections between individuals and within groups.

Now we examine pros and cons of FTF communications. The positives begin with the fact that this communication environment is about the sender making a personal connection with the receiver. Face-to-face communications allow us to get indications about tone and demeanor, and it allows us to deliver messages containing bad news, emotional meanings, apologies, and affections. We can and should deliver these messages in this human way because it makes us accountable, lets the person know who is delivering the information, and allows us to engage in managing the stress of the communication. Senders can do this because they have access to facial expressions, eye contact, tone, pitch, speed, and body language during the communication. These are fully human interactions that are sometimes time consuming, but they allow us to show real compassion.

The cons of FTF interactions are basically the reverse of the online conditions. These messages are time consuming and senders don't always think about what they want to say before they say it, which can be a result time constraints or the stress of the communication. Depending on the setting, this communication method can make it hard to edit or revise what we need to say before we say it. Another potential difficulty in FTF communication is that the receiver can respond right away. This can be a difficulty if the quick response creates the need for a new or clarifying message or if the receiver doesn't respond right away and creates awkward silence. The sender is faced with stress in having to determine whether to ask for a response, send a clarifying message, or end the communication. FTF conversations do not offer either party anonymity, and any messages that are shared outside of the initial communication may not be verbatim, or may be less than accurate (unless one records it or takes notes). These issues are what make this a time-consuming process. Senders are receivers are developing a relationship, but the difficulty is in deciding when and if the communication is to end and whether shared understanding has been achieved. This method makes time and distance major factors, because our communication is bound by the when and where. The last observation about FTF interaction challenges is that it is sometimes difficult to look someone in the face, take into account verbal and visual cues, and say what we mean, mean what we say, and say everything that's on our mind right now all at the same time.

FTF and online communications have good and bad features. That's why it's important to understand all aspects of both and work to achieve the most effective communication possible, regardless of the method. Use Table 1 to remind yourself of the pros and cons as you communicate in your preferred method.

Table 1. Comparing FTF and online conversations

FTF vs Online Comparison		
Message Encoding	**FTF**	**Online**
Sending Messages	Time consuming	Fast
Message Accuracy	Compose on the fly	Review before send
Receiver Response	Fast or slow	Fast
Sender Identity	Public	Anonymous
Sender-Receiver Relationship	Personal	Impersonal
Message Environment	Close proximity required	No time/distance requirement
Verbal & Visual Cues	Access to facial expressions, eye contact, tone, pitch, speed, body language, etc.	No access to facial expressions, eye contact, tone, pitch, speed, body language, etc.
Message Empathy	Deliver messages containing bad news, emotional messages, apologies, and sharing affections.	Don't deliver messages containing bad news, emotional messages, apologies, and sharing affections.

REFERENCES

Bachmann, R., Knights, D., & Sydow, J. (2001). Special Issue: Trust and control in organizational relations. *Organization Studies*, *22*(2). doi:10.1177/0170840601222007

Baek, Y. M., Wojcieszak, M., & Delli Carpini, M. X. (2012). Online versus face-to-face deliberation: Who? Why? What? With what effects? *New Media & Society*, *14*(3), 363–383. doi:10.1177/1461444811413191

Baker, W. E. (2000). *Achieving success through social capital: tapping the hidden resources in your personal and business networks*. San Francisco: Jossey-Bass.

Bandura, A. (1969). *Principles of behavior modification*. New York: Holt, Rinehart and Winston.

Bandura, A. (1977a). *Social Learning Theory*. Englewood Cliffs, NJ: Prentice-Hall.

Becerra, M., & Gupta, A. K. (2003). Perceived trustworthiness within the organization: The moderating impact of communication frequency on trustor and trustee effects. *Organization Science*, *14*(1), 32–44. doi:10.1287/orsc.14.1.32.12815

Brown, C. (2013). *Are We Becoming More Socially Awkward? An Analysis of the Relationship Between Technological Communication Use and Social Skills in College Students*. Academic Press.

Brown, M. A. (2011). *Social networking and individual performance: Examining predictors of participation*. Old Dominion University.

Carter, M. J. & Fuller, C. (2015). *Symbolic interactionism*. Academic Press.

Cooper, A., & Sportolari, L. (1997). Romance in cyberspace: Understanding online attraction. *Journal of Sex Education and Therapy*, *22*(1), 7–14.

Dirks, K. T., & Ferrin, D. L. (2001). The role of trust in organizational settings. *Organization Science*, *12*(4), 450–467. doi:10.1287/orsc.12.4.450.10640

Ferrin, D. L., & Dirks, K. T. (2003). The use of rewards to increase and decrease trust: Mediating processes and differential effects. *Organization Science*, *14*(1), 18–31. doi:10.1287/orsc.14.1.18.12809

Ferrin, D. L., Dirks, K. T., & Shah, P. P. (2006). Direct and indirect effects of third-party relationships on interpersonal trust. *The Journal of Applied Psychology*, *91*(4), 870–883. doi:10.1037/0021-9010.91.4.870 PMID:16834511

Gambetta, D. (1988). *Trust: Making and breaking cooperative relations*. New York, NY: B. Blackwell.

Gheorghiţă, V., & Pădureţu, A. (2014). Social Networks and Interpersonal Communication. *System, 2*(3), 5.

Goldstein, A. P., & Sorcher, M. (1974). *Changing Supervisory Behavior*. New York: Pergammon.

Grey, C., & Garsten, C. (2001). Trust, Control and Post-Bureaucracy. *Organization Studies*, *22*(2), 229–250. doi:10.1177/0170840601222003

Gunawardena, C. N. (1995). Social presence theory and implications for interaction and collaborative learning in computer conferences. *International Journal of Educational Telecommunications*, *1*(2), 147–166.

Gursakal, N., Oguzlar, A., Aydin, Z. B., & Tuzunturk, S. (2009). Measuring trust in an intra-organisational context using Social Network Analysis. *International Journal of Management and Enterprise Development*, *6*(4), 494–512. doi:10.1504/IJMED.2009.024238

Hatala, J.-P., & Fleming, P. R. (2007). Making Transfer Climate Visible: Utilizing Social Network Analysis to Facilitate the Transfer of Training. *Human Resource Development Review*, *6*(1), 33–63. doi:10.1177/1534484306297116

Huff, L., & Kelley, L. (2003). Levels of organizational trust in individualist versus collectivist societies: A seven-nation study. *Organization Science*, *14*(1), 81–90. doi:10.1287/orsc.14.1.81.12807

Ivancevich, J. M., Konopaske, R., & Matteson, M. T. (2008). Organizational Behavior and Management. New York: McGraw-Hill/Irwin.

Jaggars, S. S. (2014). Choosing Between Online and Face-to-Face Courses: Community College Student Voices. *American Journal of Distance Education*, *28*(1), 27–38. doi:10.1080/08923647.2014.867697

Kramer, R. M. (1996). *Trust in organizations: Frontiers of theory and research*. Thousand Oaks, CA: Sage Publications.

Lambright, K. T., Mischen, P. A., & Laramee, C. B. (2010). Building Trust in Public and Nonprofit Networks: Personal, Dyadic, and Third-Party Influences. *American Review of Public Administration*, *40*(1), 64–82. doi:10.1177/0275074008329426

Lane, C., & Bachmann, R. (1998). *Trust within and between organizations: Conceptual issues and empirical applications*. New York: Oxford University Press.

Li, F., & Betts, S. C. (2003). Between Expectation and Behavioral Intent: A Model of Trust. In *Proceedings of the Allied Academies International Conference*. Academy of Organizational Culture, Communications and Conflict.

Luhmann, N. (1988). *Familiarity, confidence, trust: Problems and alternatives. Trust: Making and breaking cooperative relations.* New York: Basil Blackwell.

Ma, L. (2007). *"I'll trust you if I expect you to trust me": An analysis of interpersonal trust, friends, and social interactions within social networks.* Tufts University.

Manz, C. C., & Sims, H. P. Jr. (1981). Vicarious Learning: The Influence of Modeling on Organizational Behavior. *Academy of Management Review*, *6*(1), 105–113.

Mayer, R. C., Davis, J. H., & Schoorman, F. D. (1995). An Integrative Model of Organizational Trust. *Academy of Management Review*, *20*(3), 709–734.

McEvily, B., Perrone, V., & Zaheer, A. (2003). Introduction to the special issue on trust in an organizational context. *Organization Science*, *14*(1), 1–4. doi:10.1287/orsc.14.1.1.12812

Newell, S., & Swan, J. (2000). Trust and inter-organizational networking. *Human Relations*, *53*(10), 1287–1328.

Olson, G. M., & Olson, J. S. (2000). Distance matters. *Human-Computer Interaction*, *15*(2), 139–178. doi:10.1207/S15327051HCI1523_4

Pea, R., Nass, C., Meheula, L., Rance, M., Kumar, A., Bamford, H., & Yang, S. et al. (2012). Media use, face-to-face communication, media multitasking, and social well-being among 8-to 12-year-old girls. *Developmental Psychology*, *48*(2), 327–336. doi:10.1037/a0027030 PMID:22268607

Przybylski, A. K., & Weinstein, N. (2013). Can you connect with me now? How the presence of mobile communication technology influences face-to-face conversation quality. *Journal of Social and Personal Relationships*, *30*(3), 237–246. doi:10.1177/0265407512453827

Rideout, V. J., Foehr, & Roberts. (2010). *Generation M: Media in the Lives of 8-to 18-Year-Olds.* Henry J. Kaiser Family Foundation.

Rosenbaum, B. L. (1978). New uses for behavioral modeling. *The Personnel Administrator*, 27–28.

Rousseau, D. M., Sitkin, S. B., Burt, R. S., & Camerer, C. (1998). Not so different after all: A cross-discipline view of trust. Academy of Management. *Academy of Management Review*, *23*(3), 393–404. doi:10.5465/AMR.1998.926617

Sako, M. (1992). *Prices, quality, and trust: Inter-firm relations in Britain and Japan*. Cambridge; UK: Cambridge University Press. doi:10.1017/CBO9780511520723

Shapiro, D. L., Sheppard, B. H., & Cheraskin, L. (1992). Business on a handshake. *Negotiation Journal*, *8*(4), 365–377. doi:10.1111/j.1571-9979.1992.tb00679.x

Sproull, L., & Kiesler, S. (1991). *Connections: new ways of working in the networked organization*. Cambridge, MA: MIT Press.

Stacey, E., & Wiesenberg, F. (2007). A study of face-to-face and online teaching philosophies in Canada and Australia. *International Journal of E-Learning & Distance Education*, *22*(1), 19–40.

Steinfield, C., DiMicco, J. M., Ellison, N. B., & Lampe, C. (2009). Bowling online: Social networking and social capital within the organization. *Proceedings of the Fourth International Conference on Communities and Technologies*. doi:10.1145/1556460.1556496

van de Bunt, G. G., Wittek, R. P. M., & de Klepper, M. C. (2005). The Evolution of Intra-Organizational Trust Networks: The Case of a German Paper Factory: An Empirical Test of Six Trust Mechanisms. *International Sociology*, *20*(3), 339–369. doi:10.1177/0268580905055480

Weiss, H. M. (1978). Social learning of work values in organizations. *The Journal of Applied Psychology*, *63*(6), 711–718. doi:10.1037/0021-9010.63.6.711

Whitty, M., & Gavin, J. (2001). Age/Sex/Location: Uncovering the Social Cues in the Development of Online Relationships. *Cyberpsychology & Behavior*, *4*(5), 623–630. doi:10.1089/109493101753235223 PMID:11725656

Zucker, L. G. (1986). *Production of trust: Institutional sources of economic structure. Research in organisational behaviour*. Greenwich, CT: JAI.

Chapter 2
The Departure from Human Interaction

OVERVIEW

The explosion of social media and digital communication was driven by technological advances, so we begin this chapter with a short journey through that history. The phone and the tablet dominate the way we communicate in the present day. This chapter covers computer mediated communication (CMC) as the term for using information technology to interact with others. CMC is popular because it's portable and allows us to communicate regardless of the limitations of geographical distance and time. Experts say that those who maintain long-distance friendships rate face-to-face communication fourth in importance behind telephone, email, and instant messaging. The evolution of cell phones and tablets has happened at breakneck speed. The smartphone is now a mega-functional, somewhat pocket-sized computer that includes a mobile app for virtually all of the capabilities that we used to have only on our desktop or laptop computer. Experts look at two related developments: mobile-friendly design and mobile-app development. Digital communicators can now use their smartphones to access the web quickly and effectively thanks to the continuing development of high-speed, large-bandwidth mobile networks (3G, 4G, Wi-Fi). An IT intelligence report yields fascinating information about the unbridled growth of technology and online communication. Half of the world's population now has a mobile subscription—up from just one in five 10 years ago, and an additional one billion subscribers are predicted by 2020, taking the global penetration rate to approximately 60%. This increase in online communication benefits brings some unfortunate news.

DOI: 10.4018/978-1-5225-1897-6.ch002

CMC typically results in a negative effect on face-to-face interactions with strangers, acquaintances, and families. We embark on this learning journey with a goal of getting online communicators to see, understand, and accept the skills and benefits of face-to-face communication. Change is necessary to get digital communicators interested in face-to-face interactions, so we turn to motivation. Motivation is the art of getting people to do something they may not otherwise want or choose to do. Online communication is also multicultural, so motivational ability must also focus on providing an effective framework for improvement and promoting heightened cultural awareness. A person's cultural background can influence his or her perceptions of the environment, motives, and intentions of behaviors, communication norms, stereotyping, ethnocentrism, and prejudices. Social cohesion and interaction can suffer if these issues are not acknowledged or addressed. The ideas about motivation, culture, and diversity presented in this chapter are focused on changing the way people think about communication by motivating them and by understanding their self-concept.

Bridging the Communication Gap

Social media and digital communication have taken over the world. But what happened to face-to-face communication? What happened to human interaction?

Technology has grown out of control over the past few decades. Gone is the time when we would gather in a room and talk together as the primary mode of communication. Today, the phone and the tablet dominate the way we communicate. Look around you while you're driving through town, on the highway, or at a stop light. How many of the people around you are looking at an electronic device? They might be texting, reading e-mail, or searching for the best location for lunch, but they're not fully watching the road and they're certainly not involved in face-to-face communication. Now you're at a restaurant and you see a table of 5 or 6 people together. None of them are looking at each other. Their eyes are on their phones, and the waitress is trying to get their attention about their order. Technology has their undivided attention.

Let me tell you about what might happen in someone's typical day when technology drives everything that happens. The alarm clock on Joe's phone rings at 5:30 a.m. He has a big meeting today, so he has to get up and get going. Joe connects his phone to a Bluetooth speaker so he can listen to his favorite tunes while he's showering and shaving. As he puts on his suit, he gets a weather alert from his phone telling him that a storm is moving out to sea, so his weekend plans may work after all. Joe gets his coffee and opens a news site on his tablet. While he's watching news, the tablet sends a traffic alert for his route to work.

Joe is all ready for work and heads for the car. The meeting is at a location he's not familiar with, so he pulls up MapQuest on his phone to get directions. His phone connects music to his vehicle for his drive to the meeting. The phone rings, suspending the music, because Joe has a call. One of his colleagues is going to the same meeting and wants a ride. Joe says he has time, so he reprograms MapQuest to stop at his co-worker's location on the way. Joe picks up his colleague Larry and inquires about how the baseball game went the day before. But Larry isn't listening because he's playing a game on his phone with a Bluetooth earpiece on.

They arrive at the meeting and Joe shuts down his music. He puts the phone on vibrate. He meets Mary, who is running the meeting. Mary lets him know this is a Skype call and there will only be 4 total people in the room. The call is connected and the meeting starts. Joe and the other three people in the room are monitoring their phones during the meeting. A person on the phone asks Larry a question, but he misses it while working on his phone. The meeting goes well and everyone is ready to head out to lunch. Mary asks where they can go, and they all check Google for "food near me." At lunch, there is some conversation, but very little eye contact.

After lunch, Joe drops off Larry. Now Joe has to go back to his house to get his phone charger because his battery is low and he forgot to bring it. We can keep going, but the idea should be clear by now. Today, technology in the form of phones and tablets is getting the attention that humans used to get. This story is typical of many situations we see today. Today, many people would rather text than talk, and they prefer mediated communication to FTF interaction.

Computer mediated communication (CMC), also known as mediated communication, mediated discourse, or mediated interaction, uses information technology to interact with others. CMC does not employ the senses or facial expressions or symbolic cues. The portable nature of CMC and its lack of distance and time limitations have even caused changes in our romantic relationships. Experts say that those who maintain long-distance friendships rate face-to-face communication fourth in importance behind telephone, email, and instant messaging (Wright & Webb 2011). As the importance of FTF communication lags behind telephone, email and instant messaging, increased digital access causes the real change from face-to-face to CMC. Technology has been around for many years, but the size and price of cell phones and tablets have made them progressively more accessible to many more people. Let's look at some history.

The demand for and accessibility of cell phones and tablets have grown dramatically over the years. There has been remarkable movement since the year 2000, with changes upon changes. Today's smartphone has evolved from a large semi-functional phone, to a smaller semi-functional phone, to a smaller fully functional phone, to a larger fully functional phone. As phones get larger to provide more screen size for watching videos, tablets tend to get larger as well. Now, the various uses and ease of handling in terms of size or weight gives us increasingly more capability and convenience every day.

Mobile-friendly design and mobile-app development brought users countless mobile applications from which to choose as the smartphone marketplace continued to grow out of control. There is virtually nothing you can't get from the marketplace and, if something is missing, you typically hear people say "We need an app!" Digital communicators continue to get increased bandwidth and reduced limitations for use, leading many to say the smartphone is just as good and capable as a desktop computer.

"Today, the mobile phone market is outstripping ownership and use of landline phones in virtually every country on Earth, the steady conversion of more and more of these phones into smarter, more multimedia-friendly devices that work off Wi-Fi, 3G, or 4G networks suggests that the future of mobile development and design will only grow stronger" (Reagle, 2012).

People used simple wireless analog-based (1G) portable phones through the 1990s when mobile phones changed in sophistication. The technology continued to improve and numerous features were developed, and those features were much more complex than anything that had been imagined before. The evolution started with our desires to have a phone that had great communication accessibility and that gave us mobility to go anywhere without being restricted by coverage area. The next development was Europe's Global System for Mobile communication (GSM), a digital mobile telephony system that was also used in other parts of the world. Sometimes called 2G, GSM allowed a huge leap in available mobile services during the mid to late 1990s.

As the pre-smartphone devices came on the scene, data transmission changed from analog to digital, allowing the devices to handle the earliest smartphone features. Those features led to today's explosion of app development. Consumers started to get downloadable content, text messaging, and very low-level web access. This basic access let consumers send emails, look at limited selections of online multimedia, and download music files, ringtones, and other simple digital applications. Advances in 2G technology convinced many to give up their landline communication tools in favor of the new capabilities in the mobile phone. These were still relatively basic devices, but that didn't stop widespread popularity worldwide that reached all income levels.

The 2000s brought the first 3G wireless digital networks with more sophisticated development and service offerings. These true smartphones were amazing, and in 2002 and 2003 network operators offered widespread 3G access.. More powerful transmission technology took over for the 2G networks circuit switching mechanism. This was unprecedented wireless transmission technology that depended on efficient packet switching data transmission. This began the modern era of wireless mobile smartphones as pocket-sized computers. Phone networks were now busy offering web-based data options for smartphones, and these offerings exploded from small quantities into out of control development. The growing app market drove the redesign of the devices to better display interactive systems and digital media.

The more powerful 4G network with packet-switching based data optimization was next. Some digital experts characterize the increases as "10-fold," based on improved data transmission ability and further improved access to digital media. The demand for media-rich mobile applications also increased again. Around this time in the overall development, the touchscreen phone with large visual screen interface was created. This technology virtually replaced small display screens and buttons. The replacement of technology was so rapid that in some countries it's no longer possible to purchase anything older than 3G. Almost all new phones today feature buttonless touchscreen design.

Reading the GSMA Intelligence report "The Mobile Economy" yields fascinating information about the unbridled growth of technology and, in turn, online communication. "The mobile industry continues to scale rapidly, with a total of 3.6 billion unique mobile subscribers at the end of 2014. Half of the world's population now has a mobile subscription—up from just one in five 10 years ago. An additional one billion subscribers are predicted by 2020, taking the global penetration rate to approximately 60%. There were 7.1 billion global SIM connections at the end of 2014, and a further 243 million machine-to-machine (M2M) connections" (GSMA, 2016).

The technology facilitates online communication. Tablets are growing bigger, faster, and lighter, just like the smartphone. There are three forms of mediated communication: mediated interpersonal communication, interactive communication, and mass communication (Lundby, 2009). This broad spectrum of facilitated communication demonstrates social media's effect on our interaction and collaboration. We know that people are building social capital and seeking social identity online leading to more interactive communication. That has changed the basic style of communicating and has continued to limit and hinder FTF communication. Some experts say that people are more interactive than before, but it comes with a price. Digital conversations create weak ties characterized by less personal feeling than we have when we're face-to-face with our receiver. Still, communicators tend to send open and honest messages because we tend to trust the people on the other end, even if we haven't met them. Because of weaker ties, it is harder to advance relationships in online communication. Also, diversity of communication may suffer because in the online environment, we basically interact with people who agree with our points of view. People ride the wave of new technology without considering the implications for closer relationships.

One research project took a broad look at technology and communication, analyzing previous studies, conducting field observations, and completing an online survey. This research was intended to determine the level of engagement that individuals have with their cell phones, with other technologies, and with each other in face-to-face situations. The study findings showed a negative effect on the quality and quantity of face to face communication. The report of the study shows that the U.S. Census reports that 76% of households reported having a computer in 2011, compared with only 8% in 1984 (Drago, 2015). The report went on to list that the number of households reporting that they accessed the internet rose from 18% in 1998 to 72% in 2011. Also, in 2013 90% of American adults had a cell phone of some kind and that number was closer to 97% for people under age 44. There appears to be no credible, similar research at this depth that addresses tablets, but we can be sure some of those numbers are included. We can definitely be sure that the conversations that are conducted via tablet are at least as busy as those of the smartphone. Also, the new compatibility between smartphones, tablets, and laptops allows us to sync all of those devices and facilitates more online communication and activity.

What these reports and others tell us is that billions of people can easily connect with others short distances or millions of miles away. The prevalence of these devices influences the departure from FTF interactions. A 2011 study highlighted mobile communication and its influence on the extent to which people engage face-to-face with new people in public settings (Campbell & Kwak, 2011). The study examined different types of cell phone uses and found that mobile phone use actually enhanced information exchange with strangers. In another study, researchers demonstrated that the presence of mobile communication devices in social settings causes problems in human relationships (Przybylski & Weinstein, 2013). In conducting two experiments, the researchers uncovered evidence that the devices negatively influence closeness, connection, and conversation quality. The research concluded that the negative influence increases when people discuss personally meaningful topics.

The relationship between mobile device availability and the quality of in-person social interactions was the subject of another study, one in which researchers discovered that conversations without the use of mobile communication technologies were rated as significantly superior to those in the presence of a mobile device (Misra, Cheng et al. 2016). People who had to communicate without the benefit of mobile devices said they had higher levels of empathetic concern, while those who used a mobile device reported lower levels (Misra, Cheng et al. 2016). These and other studies provide evidence that mobile phone use actually facilitated talking with strangers to get and exchange information. There were some findings that indicated that "cyber-youth" have grown up with cell phones and the internet in education, communication, and entertainment, resulting in a significant decrease in face-to-face interaction among this group. All of this information about online activity, identity, and communication points to the reality that there is some level of degradation in the social skills and the presentation of self-identity to others. There are indications that there are consequences for personal relationships as well.

> *In* Alone Together: Why We Expect More from Technology and Less From Each Other, *Sherry Turkle (2012) examined the effects of technology on familial relationships. After interviewing more than 300 young people and 150 adults, Turkle found that children were often times the ones complaining about their parents' obsession with technology. Turkle discovered that many children believed their parents paid less attention to them than to their smartphones, often times neglecting to interact with them face to face until they had finished responding to emails. (Drago, 2015)*

The evidence above tells us that while there are great benefits to online communication those benefits also costs us. CMC typically results in a negative effect on face-to-face interactions with strangers, acquaintances, and families. Online communication puts people in touch and helps them manage daily activities. The interesting part is that it doesn't create strong, long-lasting relationships. It creates weak ties, as we said earlier. We sacrifice connection and empathy for speed and ready access. This is anytime, anywhere connectivity that allows communication in a variety of ways. The very public use of this technology causes issues. It is interesting that we can connect with anyone, anywhere, but when using the technology in public, we don't connect with the people sitting right next to us. This is true even if those people are family and friends. We are committed to the technology.

Public use challenges behavioral norms, and many people are still uneasy or irritated when someone else uses technology in public. This is most often a concern when people are taking and participating in calls. This bothers copresent others. Copresence can be defined as the perception by a communicator that another person in a mediated or online environment is real, immediate, or present. The mobile communicator can be oblivious to the objection around him or her. This is interesting, because it is the basis for many of the arguments against sole reliance on digital communication in this book. Human interaction is necessary to keep a sense of society and build real relationships.

Motivation and Culture Considerations

As we embark on a learning journey with a goal of getting online communicators to see, understand, and accept the skills and benefits of face-to-face communication, we have an awesome challenge. The digital conversation is going very well, and full-range communication is not on the radar as an issue of concern. This book will attempt to change that thought process. So why are we defining motivation? Motivation is the art of getting people to do something they may not otherwise want or choose to do. Those who are great at motivating can get people to exhibit certain behaviors without knowing that it's happening. It will take motivation to change the minds of online communicators who are not taking advantage of face-to-face communication as part of the interaction skill set.

So motivation consists of internal and external factors that stimulate the desire and energy for people to be continually interested and committed to a job, role, or subject, or to strive for a goal. Conscious and unconscious factors come together to create motivation, and those factors are:

1. Intense desire or need,
2. A goal that offers incentive or a reward, and
3. Expectations of the individual and his or her peers.

Online communication is also multicultural, so an additional focus of motivational ability is to provide an effective framework for improvement and to promote heightened cultural awareness (Shenkar & Zeira, 1992; Marquardt & Horvath, 2001). There is a lack of research to support various ways to motivate, because experts have shown an increasing failure to focus on motivation, looking instead at human performance or other behavioral factors.

This analysis is about motivation and how it may be affected by the cultural diversity we see in online communication. We will address deficiencies in motivating by offering several propositions that illustrate the failure to properly manage critical success factors. This failure can hinder motivational activities. Attending to success factors can identify weaknesses that many communicators face. By identifying these problem areas, we can work to strengthen skills and, in turn, improve our ability to motivate.

In addition to assistance in developing a motivational plan, culture is important because diversity is on the rise in our society. The U.S. Census Bureau 2000 reported that the foreign-born population in the U.S. increased by more than half between 1990 and 2000 (Malone 2003). The foreign-born population increased 57 percent, from 19.8 million to 31.1 million, during that time. By 2002, the bureau reported that the estimated size of the foreign-born population was 32.5 million (Schmidley, 2003). These examples illustrate that the cultural makeup of America continues to shift. Communicators should consider this. America's multiethnic composition requires that we pay attention to culture, because a person's cultural position is a good indicator of how he or she makes decisions.

The first step in improving motivational ability is gaining an understanding of multicultural situations. Smith and Luckasson (1992) found that more than half of the students in 25 of the largest cities in the U.S. are from diverse groups in terms of linguistics, ethnicity, and cultural background. Their findings show that of the 53 million children in America's elementary and secondary schools in 1992, 35 percent were of non-White origin. That figure is projected to increase 51 percent by the year 2050 (Futrell, Gomez et al., 2003). Various researchers agree about the continuing change in U.S. multiethnicity. The U.S. population is expected to increase from 284 million to 325 million by the year 2020, and the Caucasian population is expected to decrease from 76 percent to 50 percent during that timeframe with, of course, corresponding increases in all other ethnic and foreign-born groups (Bell & Hurd, 2006).

Because of demographics and the unlimited reach of the digital environment, online communicators may face multicultural situations due to issues of multiethnicity, so they must acknowledge the various cultures they are dealing with. When possible, senders should consider life experience, conflict behavior, and personality of the receiver. Motivation is important to our journey and, when we discuss group communication, team cohesion will be understood as fundamental to effective communication.

Matveev and Nelson (2004) outline factors that can be useful in improving communication. A person's cultural background can influence his or her perceptions of the environment, motives and intentions of behaviors, communication norms, stereotyping, ethnocentrism, and prejudices. Social cohesion and interaction can suffer if these issues are not acknowledged or addressed. The problem is that multicultural interactions may not be available due to a pre-existing identity because few commonalities exist (Earley & Mosakowski, 2000). In this case, senders must ensure that the communication is easy to follow, including simplified rules, performance expectations, and attempts to elicit feedback. The goal is to share expectations leading to improved full-range communications.

Cultural differences in communication activities can cause conflict and misunderstanding (Shenkar and Zeira 1992). There are five widely accepted challenges relating to cultural differences:

1. Managing cultural diversity and conflicts;
2. Handling geographic distances and dispersion;
3. Dealing with coordination and control issues;
4. Maintaining communication richness; and
5. Developing and maintaining team cohesiveness (Marquardt & Horvath 2001).

Knowledge of culture and language norms is a first step in improvement. Success also requires skills in the areas of human warmth, charisma, empathy, and the ability to manage anxiety and uncertainty (Gudykunst 1998).

At a basic level, senders can assess players' individualism-collectivism orientation based on Hofstede's principles (1991). Cultural orientation gives an indication of the degree to which people of a certain culture are motivated to join or take part in societal or organizational groups. Individualistic cultures value individual goals, while collectivistic cultures value group goals. Cultures that are collective exhibit a higher level of emotional dependence on the team, and are more conforming, orderly, traditional, team-oriented, and particularistic. Individualistically oriented cultures, such as the United States, Germany, and Hungary, tend to value the concepts of autonomy, self-interest, and performance. In contrast, collective cultures, such as Japan, Sweden, and Russia, value group harmony, cooperation, and satisfaction (Matveev and Nelson 2004).

Culture is important for group cohesiveness, collaboration, and communication. Research has shown the importance of culture to recognize and assign importance to the interaction. Randel (2003) established culture as a critical factor for group members. The study examined the extent to which culture is salient to multinational team members. The research provides evidence that cultural identity salience is important to team performance through its effect on team citizenship behavior. Subjects of the study identified situations in which team members viewed cultural identity as conspicuous or prominent.

This look at cultural salience uses Randel's (2003) definition of cultural identity salience, which describes it as the extent to which the cultural identities of team members are conspicuous or prominent as assessed by an individual in that team. This definition employs, and is consistent with, the findings of McCall and Simmons (1978). Individuals use an automatic social categorization process because they want to make sense of their social environment and because there are cognitive limitations involved (Fiske, 1990).

Individuals have varying psychological experiences, and cultural identity salience allows the observer an alternative to approaches that rely on the nationality categories to which individuals belong. Individuals may get team assignments from their organization based on their cultural background with the expectation that their behaviors will be similar to the a priori predictions about national differences. This approach does not fully account for the effects of psychological experiences. Hambrick et al (1998) found that individuals may not display behaviors or attitudes representative of their nationality. This may be due to several factors: individuals may have had other international experiences, their personal experiences may differ from those of others from the same nationality, or socialization based on the company's culture may have affected employees in ways that are different from what one would predict based solely on nationality. This is important information and face-to-face communication would provide verbal and visual cues, along with feedback, to assist any analysis of our communication partner.

All of the aforementioned findings measure the extent to which individuals are aware of the cultural backgrounds of the people around them. A key to successfully employing cultural identity salience to observations is to ensure the unit of analysis is the individual. Earley (1994) showed a relationship between efficacy training and individual performance as they relate to a person's individualism-collectivism orientation. By measuring orientations of people from China, Hong Kong, and the U.S. rather than assuming their individualism-collectivism orientation, Earley kept the predictor and the dependent variables at the same analytical level – the individual. This approach allows for individual variability in the endorsement of cultural values. This kind of emphasis is perfectly suited for a high-touch approach to relationship-building in communications.

Randel (2003) cautions that the results suggest that the extent to which an individual has the same country of origin as other team members influences cultural identity salience. At least two possible scenarios can be at work in this area. An individual who holds a particular cultural value may not behave in ways that reflect that cultural value. Alternatively, members of a multinational team might assimilate to each other so that their cultural differences do not appear to play a role in interactions among team members.

What this all means is that culture is ever present and it is always important to be aware of its influence. Culture and the ways people react to it exerts a powerful influence on organizations. When culture is conspicuous, it can be a benefit to team cohesion, because team members cannot ignore how they differ from one another. Understanding cultural orientations is the right way to begin any motivating effort.

As senders strive to understand others, they can focus on five cultural orientations: richness of the communication context, power distance, individualism, uncertainty avoidance, and performance orientations (Matveev & Nelson, 2004). Studying the role of communication in culture, and distinguishing cultures and communication by the information surrounding an event regardless of the verbal message, Hall (1976, 1989) viewed cultures on a low- to high-control continuum. Low levels of programmed information are used in low-context cultures. In this instance, words are an explicit code that carries a message that includes the medium, bits of information, and explicit meaning. Marquardt and Horvath (2001) maintain that North America's task-centered communication is a good example of low-context communication, characterized by transactions that occur with relatively little information exchanged about a person or a company. Physical setting and an individual's internalized values, beliefs, and norms convey the message in a non-verbal context for high-context cultures (1976, Hall, 1989). For instance, in Russian high-context culture, meaning is shared implicitly because the listener knows the context and needs little background information (Hall, 1989). Communication and behavioral rules are implicit in the context and communicators need rich contextual information about a person or a company before business transactions can be completed successfully (Marquardt & Horvath 2001).

The preceding findings show that senders need to know what factors hinder their communication and motivation effort, and what tools they have at their disposal. A lack of knowledge of other cultures in terms of norms and values can cause problems not only in motivation, but also in group communication, rewards, conflict resolution, and other issues critical to successfully motivating teams. Each group member's individualistic or collectivistic position has an impact on response to motivation and possibly on willingness to interact.

When dealing with multiethnicity, there can be problems with motivation for two reasons:

1. Failure to fully understand motivation; and
2. Lack of understanding of the cultures and sub-cultures the coach deals with.

It is necessary to address the motivation issue before moving on to the cultural understanding issue.

Researchers have generally abandoned the concept of motivation, focusing instead on human behavior. Ambrose and Kulik (1999) emphasize the lack of results in this area, finding that researchers have not been studying motivation. Motivation research has been replaced by the specific pursuit of employee behavior, such as in the areas of task performance, organizational citizenship behaviors, and group communications. Researchers have not been measuring motivation; they have been using motivational theories to predict or to suggest that motivation is occurring. Ambrose and Kulik (1999) wrote:

They had other, more specific goals in mind, such as studying employee performance in work teams, or studying when employees engage in discretionary behaviors.... Established motivational theories were an effective framework for predicting these behaviors when researchers did not explicitly measure "motivation," or even suggest 'motivation' as an explanatory mediating construct. (p. 279)

These are exactly the reasons motivation, culture, and multiethnicity are included in our analysis of communication effectiveness. These are all relevant factors in improving communication with people and groups. As we move through this learning journey, information will be provided on flexibility in communication. In other words, the more skills one has that assist or support the communication activity, the better the chances for successful outcomes.

Defining Motivation

Defining important terms provides the basis for a motivational approach to communication. Again, motivation is the degree to which an individual wants and chooses to engage in certain specified behaviors (Mitchell, 1982). There are many definitions of motivation. The nonacademic definition is the degree to which an individual wants and tries hard to do well at a particular task or job (Mitchell, 1982). The dictionary describes motivation as an impetus to action. Social scientists suggest a more technical definition where motivation refers to the psychological processes that cause the arousal, direction, and persistence of behavior (Atkinson, 1964; Campbell, Dunnette et al., 1970; Korman, 1974; Huse & Bowditch, 1977; Luthans, 1977; Kast & Rosenzweig, 1979). The definition can also include emphasis on goal direction or voluntary aspects. So, motivation can be viewed as psychological processes that cause the arousal, direction, and persistence of voluntary actions that are goal directed (Mitchell, 1982).

Mitchell (1982) thus provides a clear definition of motivation that is appropriate for this project. The researcher says that motivation is the degree to which an individual wants and chooses to engage in certain specified behaviors, and contends that there is consensus about some of the underlying properties of motivation. Motivation is an individual phenomenon, it is intentional, it is multifaceted, and it can predict behavior.

Defining Cultural Intelligence

A definition of cultural intelligence is also provided because the construct is valuable to our learning experience. According to Lévy-Leboyer (2004) in *Cultural Intelligence: Individual Interactions across Cultures*, cultural intelligence has three components: cognitive, motivational, and behavioral. Cultural intelligence is the ability to understand other cultures based on cognitive (conceptual understanding), motivational (process and components), and behavioral (engage in adaptive behaviors) learning. (Lévy-Leboyer, 2004)

The cognitive aspect refers to the skills needed to conceptualize a new culture and to gather information about a new world. The motivational component deals with the desire to adapt to the other culture, rather than only seeing it as different. Understanding the motivational aspect requires a deep understanding of the motivation process and of its components, such as self-efficacy, self-concept, and goal setting. Understanding others' culture and wishing to adapt to it is very important. The behavioral component is the "capability of an individual to actually engage in behaviors which are adaptive." (Lévy-Leboyer, 2004)

Research shows cultural intelligence as a model with three distinct elements. To assess personal cultural intelligence, one would examine:

1. The understanding and knowledge of different cultures;
2. Willingness to address different cultures and respond appropriately; and

3. Ability to change personal behavior to adjust to what other cultures require.

How successfully senders manage these elements is an indication of cultural intelligence.

Improved information on motivation and cultural intelligence assists with communication activities. Further, there are five critical areas where people may have difficulty when motivating people:

1. Management of interaction adjustment norms;
2. Creation of mutual adjustment processes;
3. Ability to improve cross-cultural communication competence;
4. Establishment of commonalities of purpose; and
5. Design of infrastructure practices that are supportive of team culture.

For instance, failure to manage interaction adjustment norms could cause someone to lose "face" because the sender has ignored or misread their position. Similarly, using the wrong language or phrase can be disastrous if it violates a cultural norm when translated. Next, unless both sides make the necessary adjustments, a group can be pulled apart in different directions and there will be no social cohesion. Difficult rules, negative perceptions, and vague performance expectations prevent groups from establishing commonalities of purpose. Any one of the aforementioned problems could cause the failure of the motivation effort. Communicators must develop solid methods for dealing with these factors.

Additionally, these factors are crucial to the motivation effort because they are specifically linked to major components of the change process. Why mention the change process? The high-teach solutions to bridging the communication gap require change; change in the way people communicate and change in the way they use technology. That's why these factors are very important to emotional considerations of the change process. The effect of emotion on radical change dynamics can be best understood by looking at the change process in separate components. There are three critical steps required to achieve radical change: receptivity, mobilization, and learning (Huy, 1999). Receptivity is a person's willingness to consider change and it is characterized as both a state and a process. At any fixed point in time, a person can accept the need for the proposed change if there is an interpretive, attitudinal state on the cognitive and emotional level.

The concrete action a person takes in the direction of change is mobilization. This is the process of rallying and propelling different segments of the organization to undertake joint action and to realize common change goals (Huy, 1999). Receptivity and mobilization are linked to, or lead to, learning. Individuals learn by thinking and then acting, using the outcome of action to revise their belief systems (Kim, 1993). Receptivity is an observed change, where individuals exhibit various stages of willingness to accept the proposed change, from resigned, passive acceptance to enthusiastic endorsement (Huy, 1999).

Huy's (1999) characterization of the radical change process maintains that when receptivity leads to motivation, individuals and organizations also can learn from the outcomes of the changes they enact, and this learning provides a feedback loop from the outcomes of behavioral change back to receptivity. In other words, the learning process is a beginning, but that beginning leads back to using that process to sustain receptivity at the desired level. In turn, sustained receptivity at the correct level leads to continued mobilization and so forth. The key issue, then, is how the radical change process begins and continues.

Receptivity depends on effective management of interaction adjustment norms, as well as on the need to put in place mutual adjustment processes. In terms of interaction adjustment, there are many forms of language, but the coach must find a way to master the language necessary for communicating with each member of the group, regardless of their cultural orientation. Assimilation and integration are useful in this endeavor. Assimilation requires that the sender learn a little of the receiver's language and use it correctly. A good example of this practice can be found in a study by Zimmermann et al. (2003) which examined the adjustment of German expatriates to intercultural interactions at work and at home. The study found that assimilation and integration were at work when expatriates learned the language and uttered criticisms more politely. Further, members of every culture have ideas about and reactions to issues of pride, shame, and reputation, or what is colloquially called call "face" (Zimmermann, Homan et al. 2003). Communicators must be able to protect each person's position in terms of "face." Additional interaction adjustment norms include developing an indirect communication style and personalizing work relationships. Indirect communication would be to understand that two different cultures could have two different views on what "I will come to dinner around 8" means. An individualistic approach might be to arrive between 8 and 9. A collectivistic approach would be to arrive 5 minutes early or right on time. Indirect communication requires understanding that there is a difference. Finally, personalizing group relationships means treating each person equally.

Instituting mutual adjustment processes is the other requirement for achieving receptivity. Norms of interaction of other cultures should be learned and integrated with one's own norms. However, it is imperative that there is accountability on the part of all members. All communication partners have equal responsibility to adjust, assimilate, or integrate. Of course, the "leader" should have mechanisms in place to help the process along, but each member must be responsive to the overall effort and be individually cooperative.

Achieving mobilization requires improving cross-cultural communication competence and establishing commonalities of purpose. The best way to accomplish this is by interacting, questioning, and sharing information. Senders must find a way to understand various personalities, conflict behaviors, and life experiences. Equally important are commonalities of purpose. This means simplifying rules, creating specific performance expectations or objectives, and understanding member perceptions. Once members can see the goal and accept it as their own, the mechanisms for moving toward goal fulfillment must be evident, consistent, and easily understood.

The learning process begins once receptivity leads to mobilization, and requires infrastructure practices that support team culture. Senders provide support to receive and share a vision of interactive communications to achieve the level of success required. Senders can offer educational or training messages to ensure the interaction achieves the agreed upon end state. All activities must be supportive of interactions that foster individual and group cohesion because, as stated earlier, cohesion is vital to overall success. Everyone involved in the communication must be committed to the effort to ensure all energy is focused on the same end state.

Finally, the lack of dedicated motivation research poses real problems. Experts agree that motivation has been neglected by scholars who have decided to take a back-door approach, choosing to focus on a variety of factors of motivation. For example, much research is based on social cohesion, social interaction, team communication, organizational citizenship behaviors, task performance, and employee behavior, to name a few. None of these approaches represents the study of motivation. That's why it is imperative to identify critical success factors that senders may have difficulty mastering, because these are at the heart of the difficulty in successful motivation. People have struggled with motivation due to a lack of knowledge of the decision patterns of others, and issues of diversity and differing cultural backgrounds can make the challenge even harder. According to current research, responding to internal and external stimuli motivates people.

The ideas on motivation, culture, and diversity presented here can lead to new techniques for achieving successful communication in a multiethnic setting. Changing the way people think about communication by motivating them and by understanding their self-concept can go a long way towards effective communication. The major benefit is that these skills help senders get the right message with the right emphasis in the right venue.

REFERENCES

Ambrose, M. L., & Kulik, C. T. (1999). Old Friends, New Faces: Motivation Research in the 1990s. *Journal of Management*, *25*(3), 231–292. doi:10.1177/014920639902500302

Atkinson, J. W. (1964). *An Introduction to Motivation*. Princeton, NJ: Van Nostrand.

Bell, C. M., & Hurd, A. R. (2006). Research Update: Recreation Across Ethnicity. *Parks & Recreation*, *41*(10), 28–34.

Campbell, J. P., Dunnette, M. D., Lawler, E. E. III, & Weick, K. E. Jr. (1970). *Managerial Behavior, Performance and Effectiveness*. New York: McGraw-Hill.

Campbell, S. W., & Kwak, N. (2011). Mobile Communication and Civil Society: Linking Patterns and Places of Use to Engagement with Others in Public. *Human Communication Research*, *37*(2), 207–222. doi:10.1111/j.1468-2958.2010.01399.x

Drago, E. (2015). The Effect of Technology on Face-to-Face Communication. *Elon Journal of Undergraduate Research in Communications, 6*(1).

Earley, P. C. (1994). Self or Group? Cultural Effects of Training on Self-Efficacy and Performance. *Administrative Science Quarterly*, *39*(1), 89–117. doi:10.2307/2393495

Earley, P. C., & Mosakowski, E. (2000). Creating Hybrid Team Cultures: An Empirical Test of Transnational Team Functioning. *Academy of Management Journal*, *43*(1), 26–49. doi:10.2307/1556384

Fiske, S. T., & Neuberg, S. L. (1990). A Continuum of Impression Formation from Category-based to Individuating Processes: Influences of Information and Motivation on Attention and Interpretation. *Advances in Experimental Social Psychology*, *23*, 1–74. doi:10.1016/S0065-2601(08)60317-2

Futrell, M. H., Gomez, J., & Bedden, D. (2003). Teaching the Children of a New America. *The Challenge of Diversity*, *84*(5), 381.

GSMA. (2016). London, UK: GSMA Mobile Economy.

Gudykunst, W. B. (1998). Applying Anxiety/Uncertainty Management (AUM) Theory to Intercultural Adjustment Training. *International Journal of Intercultural Relations*, *22*(2), 227–250. doi:10.1016/S0147-1767(98)00005-4

Hall, E. T. (1976). *Beyond Culture*. Garden City, NJ: Anchor Press.

Hall, E. T. (1989). *The Dance of Life: The Other Dimension of Time*. New York: Doubleday.

Hambrick, D. C., Davison, S. C., Snell, S., & Snow, C. C. (1998). When Teams Consist of Multiple Nationalities: Towards a New Understanding of the Implications. *Organization Studies*, *19*(2), 181–205. doi:10.1177/017084069801900202

Hofstede, G. (1991). *Cultures and Organizations: Software of the Mind*. London: McGraw-Hill U.K.

Huse, E. F., & Bowditch, J. L. (1977). *Behavior in Organizations: A Systems Approach to Managing*. Reading, MA: Addison-Wesley.

Huy, Q. N. (1999). Emotional Capability, Emotional Intelligence, and Radical Change. *Academy of Management Review*, *24*(2), 325–345.

Kast, E. F., & Rosenzweig, J. E. (1979). *Organization and Management: A Systems Approach*. New York: McGraw-Hill.

Kim, D. H. (1993). The link between individual learning and organizational learning. *Sloan Management Review*, *35*(Fall), 379–500.

Korman, A. K. (1974). *The Psychology of Motivation*. Englewood Cliffs, NJ: Prentice-Hall.

Lévy-Leboyer, M. (2004). Cultural Intelligence: Individual Interactions across Cultures by Christopher Early and Soon Ang. *Personnel Psychology*, *57*(3), 792–794.

Lundby, K. (2009). *Mediatization: Concept, changes, consequences*. Peter Lang.

Luthans, F. (1977). *Organizational Behavior*. New York: McGraw-Hill.

Malone, N., Baluja, K. F., Costanzo, J., & Davis, C. (2003). *The Foreign-Born Population: 2000*. Washington, DC: U.S. Census Bureau.

Marquardt, M. J., & Horvath, L. (2001). *Global Teams: How Top Multinationals Span Boundaries and Cultures with High-Speed Teamwork*. Palo Alto, CA: Davies-Black.

Matveev, A. V., & Nelson, P. E. (2004). Cross Cultural Communication Competence and Multicultural Team Performance: Perceptions of American and Russian Managers. *International Journal of Cross Cultural Management*, *4*(2), 253–270. doi:10.1177/1470595804044752

McCall, G. J., & Simmons, J. L. (1978). *Identities and Interactions: An Examination of Human Associations in Everyday Life*. New York: The Free Press.

Misra, S., Cheng, L., Genevie, J., & Yuan, M. (2016). The iPhone Effect: The Quality of In-Person Social Interactions in the Presence of Mobile Devices. *Environment and Behavior*, *48*(2), 275–298. doi:10.1177/0013916514539755

Mitchell, T. R. (1982). Motivation: New Directions for Theory, Research, and Practice. *Academy of Management Review*, *7*(1), 80–88.

Przybylski, A. K., & Weinstein, N. (2013). Can you connect with me now? How the presence of mobile communication technology influences face-to-face conversation quality. *Journal of Social and Personal Relationships*, *30*(3), 237–246. doi:10.1177/0265407512453827

Randel, A. E. (2003). The Salience of Culture in Multinational Teams and its Relation to Team Citizenship Behavior. *International Journal of Cross Cultural Management*, *3*(1), 27–44. doi:10.1177/1470595803003001848

Reagle, D. (2012). *10 Years of Mobile Industry History in 10 Minutes*. Academic Press.

Schmidley, A. D., & Robinson, J. G. (2003). *Measuring the Foreign-Born Population in the United States with the Current Population Survey*. Washington, DC: Population Division, U.S. Bureau of the Census.

Shenkar, O., & Zeira, Y. (1992). Role Conflict and Role Ambiguity of Chief Executive Officers in International Joint Ventures. *Journal of International Business Studies*, *23*(1), 55–75. doi:10.1057/palgrave.jibs.8490259

Smith, D. D., & Luckasson, R. (1992). *Introduction to Special Education*. Needham Heights, MA: Allyn & Bacon.

Turkle, S. (2012). *Alone together: Why we expect more from technology and less from each other*. Basic Books.

Wright, K. B., & Webb, L. M. (2011). *Computer-mediated communication in personal relationships*. New York: Peter Lang.

Zimmermann, A., Homan, D., & Sparrow, P. (2003). Unraveling Adjustment Mechanisms: Adjustment of German Expatriates to Intercultural Interactions, Work, and Living Conditions in the People's Republic of China. *International Journal of Cross Cultural Management*, *3*(1), 46–68.

Chapter 3
Social Networking and Communication Research

OVERVIEW

This chapter examines a range of communication theories that are relevant to our discussion. We will begin with research relevant to social networking, and then focus on communication knowledge and discovery. The relevant research on social networking is aimed at enhancing participation, which means addressing the improvements that are necessary to identify determinants and opportunities that affect levels of acceptance and participation. Participation and understanding go hand in hand, and the current literature review begins with innovation adoption (Hatala & Fleming, 2007) and the diffusion of innovations theory. We then focus on the governing by network methodology (Goldsmith & Eggers, 2004) which focuses on linking social networking challenges to constructs intended to foster behavioral changes or adjustments that improve or enhance the networking experience. The research included also demonstrates that people should be presented with value propositions that either convince them to participate at higher levels than they presently do or that break down their rejection barriers to the activity.

In terms of communication, the 1940s featured the two-step flow process, with Paul Lazarsfeld discussing that interpersonal communication with opinion leaders mediates how mass communication activities. Leon Festinger's findings that humans are sensitive to inconsistencies in actions and beliefs led to the cognitive dissonance theory, which says that recognizing such inconsistencies causes dissonance and motivates individuals to resolve it. The greater the dissonance, the more motivated people tend to be to resolve it.

DOI: 10.4018/978-1-5225-1897-6.ch003

Knowledge gap theory states that those who have greater access to the media or information acquire information faster than those whose status is lower, and that the gap tends to increase between these segments. Social presence theory, discussed by Charlotte Gunawardena, demonstrates the degree to which a person is perceived to be real in mediated communications. This theory was originally explained as the degree of salience that one person has versus the consequent salience of the interpersonal relationships.

As we examine the theories mentioned to this point, and the ones that follow, we must keep in mind that trust is an important consideration. Trusting relationships in online communications involve an ongoing decision to give most people the benefit of the doubt, and that trust is often extended to people one does not know from direct experience. This examination of trust in the communication context requires focusing on two central issues:

1. Using trust to deal with uncertainty and
2. Acceptance of vulnerability.

Trust is important because it helps us with the next theory, social context cues. We search the person's "soul" for the wealth of FTF communication that occurs without words, instead coming from tone of voice, dress, use of space, and body language. This examination also focuses on a major perspective in sociology called interactionist theory. Interactionist theory is focused on the concrete details about what happens among people in everyday life. Interactionists study how people use and interpret symbols to enhance and improve communication, and then use this information to create a sense of self and to add realism to the social situations they experience, working to create and sustain those activities. The chapter concludes with an in-depth look at the three theories that are fundamental to this examination of the communication gap between FTF and online interactions: Emotional intelligence, sensemaking, and communication accommodation theories (CAT). EI is the ability to recognize the emotions in yourself and others and to use this information to make improvements in self-management and relationships with others. Fundamentally, EI is a process of gaining awareness by recognizing personal emotions and those of others. Sensemaking theory is about social activities that require decoding messages for ourselves, drawing upon a common language we use and on the actions we engage in, and understanding the importance of sharing ideas and influencing how others make sense of events. CAT states that people adjust their speech, vocal patterns, and gestures to accommodate others when they communicate, providing a way to examine why people work to emphasize or minimize the differences between themselves and others during non-verbal and verbal communication. The theory also focuses on important interaction considerations like language, context, identity, and intergroup and interpersonal factors.

Current Research in Social Networking Theory

Much of the growing body of social networking literature examines social capital theory (Putnam, 1993) and social network analysis (Hatala, 2006). These theories are useful for the current examination of social networking. Connectivity in social networking, or innovation adoption, is about an organization's ability to get connected at an optimal performance level based on stated goals and objectives. To be successful, leadership must create an atmosphere capable of maintaining the necessary structural support mechanisms for information sharing and knowledge management (Hatala & Lutta, 2009).

There are three main components in the diffusion of innovation process leading to achievement within a social system: the innovation-decision process, the characteristics of an innovation, and adopter characteristics (Perry, 2006). The theory addresses using density levels, social structure, and demographic characteristics to manage organizational connectivity. Another theory, social network analysis (Hatala, 2006), can identify density level, which is connectivity among and between various network groups.

The governing by network approach (Goldsmith & Eggers, 2004) involves helping governments ensure their network-based partnerships are administratively effective and politically accountable.

Governments should move away from having employees view themselves as doers and instead try to create a culture where employees view themselves as facilitators, conveners, and brokers of how to engage the community's talents to accomplish the task at hand. This approach requires less reliance on public employees in traditional roles and more on a web of partnerships, contracts and alliances to do the public's work. In this way, network initiatives can now address accomplishing public objectives with measurable performance goals, assigned responsibilities to each partner, and structured information flow. The ultimate goal of these efforts is to produce the greatest possible value proposition, even greater than each participant could achieve on their own without collaborating (Brown & Schario, 2014).

It's important that leaders enhance social networking activities by paying attention to cultural compatibility when selecting network partners. This is essential for fostering long-standing, mutually beneficial relationships and for achieving shared value across the culture. Cultural compatibility creates ties that bind, leading to effective network ventures, dependable communication channels, coordination between network participants, and development of trusting relationships. The challenge is much greater than using the technology to manage relationships. It encompasses the need for attention to personnel issues, examination of processes, value creation, and trust building. Network integrators must create and maintain the infrastructure and conditions that support long-term relationship building. But network integrators are communicators first, and they must focus on network governance, knowledge sharing, value and incentive alignment, trust building, and overcoming cultural differences in equal amounts throughout their efforts.

When these responsibilities are managed effectively, participants can benefit for enormous value creation. The value is provided based on the varied and unlimited points of employee contact that can be translated into useful responses by everyone involved in the network, allowing each to adjust their responses appropriately. Governing by network is a way to address the limitations we find in social networking research.

A basic understanding of social capital theory (Putnam, 1993) and social network analysis (Hatala, 2006) is helpful for the high-teach experience in this book. Social capital addresses the benefits of relationships while social network analysis offers ways to understand how relationships are structured. Social capital theory is based on the notion that relationships matter and that social networks are a valuable asset (Field, 2003). People benefit from community-building interactions, leading to commitments that create social ties as they knit a kind of "social fabric." The resulting trust relationships are essential for social networking experiences that build strong ties and a sense of belonging.

Individuals build social capital during communication interactions, establishing relationships purposefully and employing them to generate intangible and tangible social, psychological, emotional, and economic benefits in short or long terms. Social capital has five dimensions: networks or lateral associations between individuals or groups, reciprocity and expectation, trust and risk based on assumptions, social norms, and personal and collective efficacy.

Social capital and organizational learning provide insight to knowledge transfer and perceived organizational performance. Rhodes et al (2008) examined these relationships, integrating organizational learning capability with social capital to shape a holistic knowledge sharing and management enterprise framework. An integrative model can produce a significant strategy to achieve organizational success, which indicates that these dimensions are distinct and have different effects on knowledge transfer.

Examining social capital in the online era demonstrates that existing approaches to studying social capital online have been limited by the practice of importing measurements from older, functionally different media (Williams, 2006). In fact, Chapter 4 will introduce readers to a new measure that can be used to build social capital and improve performance in social networking activities. For now, we know that Williams (2006) attempted to theorize, create, and validate a series of scales to measure social capital in online and offline contexts, finding 10-item scales that are valid and reliable. The confirmatory factor analysis used in the research was primarily concerned with bridging and bonding as two distinct but related dimensions of social capital.

Fabio Sabatini (2009) contributed a new framework for measurement of social capital and social networks and a single, synthetic measure dealing with the configuration of social capital. He also provided an empirical assessment of the relationships between different types of social capital. Prior literature is filled with various motivations for organizations to bring people together online and use social networking to integrate virtual communities (VCs) into existing business models. Companies are interested in social networking, for instance, because they gain measurable benefits, such as innovation, more effective marketing, more rapid access to information, lower costs, and higher revenues (PaloAltoNetworks 2009). In its Fall 2009 edition application and usage report, Palo Alto Networks reported that an Enterprise 2.0 survey showed that the top three business benefits of social networking identified were knowledge sharing at nearly 60%, reduced effort in information gathering (nearly 50%), and improved efficiency/speed of delivery (35%).

There is an increasing amount of information in the literature on social networking (Cerulo, 1990; Zack & McKenney, 1995; Fowler, Lawrence et al. 2004; Chung, Hossain et al. 2007), trust (Newell & Swan, 2000; Nyhan, 2000), measuring performance or productivity (Davis & Luthans, 1980; Bearman, Guynup et al. 1985; Otis, 2007; Akdere & Roberts, 2008), capacity or skill building (Shetzer, 1993; Segrest, Domke-Damonte et al. 1998; Eastin & LaRose, 2000; Erickson & Jacoby, 2003) and other constructs. One relevant area of focus for social networking is provided in a dissertation on online social networking that focuses on relationships between user characteristics and use or performance, maintaining that they are not adequately addressed (Mew, 2009). The concept of "fit" is used to determine whether there are user, task or systems characteristics that are associated with the best performance and usage levels. Online Social Networking (OSN) systems such as Myspace, Facebook, and Friendster have achieved tremendous popularity, but there has been little research done on factors motivating users to use and adopt OSN systems. Because relationships between user characteristics and use/performance are not adequately addressed, Mew (2009) focuses on how end user Computer Self-Efficacy (CSE) affects use of OSN systems.

Bandura's (1986, 1989) Social Cognitive Theory suggests that human actions are the result of interaction between environment, personal traits, and behavior. The theory defines self-efficacy as an individual's expectations regarding their ability to organize and complete tasks required to ensure specific outcomes. Bandura states, "There is a marked difference between possessing sub skills and being able to use them well under diverse circumstances. For this reason, different people with similar skills, or the same person on different occasions, may perform poorly, adequately, or extraordinarily." (Bandura, 1986)

The Mew dissertation (2009) is valuable for its examination of social networking and individual performance because it investigates voluntary use of network software and performance measurement of users in non-mandatory systems. Users are influenced by network merits, rather than by the organization's mandate, and there are important interactions between CSE, task, and OSN system characteristics.

These interactions have an effect on performance and utilization. A study of social networking and its relationship to individual performance in organizations seems to answer Mew's (2009) call for further research that poses questions about when people join social systems. That study examined who brings people in and the relevant networking and social effects of their participation.

Ma (2007) conducted an analysis of interpersonal trust, friends, and social interactions. The research contributed to existing literature on social interactions with an analysis of motivations to gain either a weak or strong tie. Finally, the computational problem of trust is to determine how much one person in the network should trust another person to whom they are not connected. This study examined determinants of an individual's choice to trust others based on an interactions-based model in which interpersonal trust is conditional based on the individual's expectation of trust by others. Ma argued for the possibility of multiple equilibria for trust within a social group and examined how individual and community characteristics might alter an individual's decision to trust.

Data from the research indicated that being disadvantaged with respect to income, educational attainment, and race tends to decrease an individual's trust of others. The adverse effect on trust indicated is an issue that must be addressed in the course of social networking. Ma (2007) explains the success of online social networking in terms of the benefit that it does not contain the same rigidities as formal organizations, but grants disadvantaged individuals, who would otherwise not have the opportunity, access to social networking ties that can be beneficial regardless of status or income.

This short review of social networking literature gives our discussion a basis for improving online communication and, in Section 3, a chance to introduce those benefits in our high-teach approach. Armed with this valuable insight, we move on to communication research.

Current Research in Communication Theory

There are many of theories important to improving communication and to the learning adventure in this book; however, our primary focus is on emotional intelligence, sensemaking, and Communication Accommodation Theory (CAT). Still, it is important to examine some other theories first. For instance, the 1940s brought knowledge about the two-step flow of communication; with functionalist Paul Lazarsfeld proposing that interpersonal communication with opinion leaders mediates how mass communication activities affect people. Lazarsfeld believed that ideas from radio and print opinion leaders flow to the less active sections of the population.

In 1957, Leon Festinger proposed that humans are sensitive to inconsistencies between actions and beliefs, leading to the cognitive dissonance theory (Cooper & Carlsmith, 2001, Harmon-Jones 2012). The theory says that recognizing this inconsistency causes dissonance and motivates individuals to resolve it. The belief is that the greater the dissonance, the more motivated the person will be to resolve it. The resolution comes in one of three ways:

1. Change beliefs
2. Change actions
3. Change perception of action (rationalization, or a change in the way people view/remember/perceive the action)

Knowledge gap theory was proposed in 1970 by Philip J Tichenor, George A. Donohue, and Clarice N. Olien. Their theory states that those who have greater access to the media or information based on their socioeconomic status acquire information at a greater rate than those whose status is lower (Chatman & Pendleton, 1995). The theory also says the gap tends to increase between these segments.

Social Presence and Social Context Cues

Charlotte Gunawardena developed a version of social presence theory to demonstrate the degree to which a person is perceived to be real in mediated communication (1995). This theory was originally explained as "the degree of salience of the other person in the interaction and the consequent salience of the interpersonal relationships." (Short, Williams et al. 1976) Some researchers argue that the concept of social presence has to important components: intimacy and immediacy (Argyle & Dean, 1965; Short, Williams et al. 1976; Weiner, 1994). Factors like physical distance, eye contact, smiling, and personal topics of conversation, for example, influence intimacy in the communication. Immediacy is a measure of the psychological distance that a sender experiences between himself or herself and the receiver. In later research, the definition of social presence was presented as a continuum (Lowenthal, 2009). One end of the continuum has a focus on interpersonal emotional connection between communicators, and the other end has a focus on whether someone is "present, there, or real." (Lowenthal, 2009) Figure 1 illustrates the importance of social presence theory for the current communication discussion.

Online communication is characterized by the belief that people are present, there, or real. The blue arrow moving toward the orange "online" oval signifies that the more the sender is able to verify the receiver as present, there, or real, the more value he or she can find in the interaction. That could lead to feedback and additional communication activities. The green arrow moving toward the orange "face-to-face" oval signifies the importance of gaining an interpersonal emotional connection at some level between sender and receiver. The communication goal is central to the interaction for both methods. The big difference is that the online communicator is ill-equipped to manage both ends of the continuum. Through visual and verbal cues, the face-to-face communicator has the tools to work at both ends of the continuum, getting answers and adjusting the interaction in real time.

Figure 1.

In online interactions, there is a level of trust that is offered, implied, and even easy. The focus is on communicating RIGHT NOW and getting my message out RIGHT NOW. Trust is an important consideration for our examination because people need a real connection to establish lasting communication relationships. Sender and receiver require a dependent relationship to engender encoding, decoding, and feedback at mature levels. Each party must be satisfied enough with the interaction to continue to communicate. Further, a commitment is required by each party to allow the collaboration and adjustments to the message delivery and receipt that make information sharing possible and valuable. There must be an extension of trust, and this is often required even though the parties have just met, and even if there is no indication the other person will behave in a beneficial way. This is the magic of communication that must happen.

Senders and receivers deal with uncertain relationships and are vulnerable based on the exchange of information; sometimes very precious information. But this exchange must happen. In fact, we communicate even when we don't intend to. So communication skills are critical. Communication is governed by an attitudinal mechanism that allows individuals to subjectively assess whether or not to expose themselves to situations where there may not be an acceptable trade-off in terms of possible damage versus received advantage. The attitude develops when individuals have accepted vulnerability to others. In online interactions, this assessment may be conscious, sub-conscious, or may not be done at all. In face-to-face interactions, however, it is necessary to continued communication. Failure to develop some form of trust could result in a quick end to the interaction. Face-to-face communications also require a level of trust, and verbal and visual cues serve to help ensure that trust is achieved.

The mention of verbal and visual cues leads us to the next relevant theory, called social context cues. Online communication is valuable, but it loses a great deal of non-verbal information. According to the social context cues theory, there is a wealth of communication that occurs face-to-face without words, instead coming from tone of voice, dress, use of space, and body language.

People adapt to communication situations using a variety social context cues. Static cues are artifacts, such as personal appearance or a seating arrangement. Dynamic cues are behaviors, like nodding, signaling, and timing. Social context cues are indicators of appropriate behavior, and they can have geographic, organizational, or situational variables. People need social cues to help them adapt to social and hierarchical needs of a situation and, lacking these cues, they are faced with more challenging communication activities in which they may be self-centered, unregulated, or unable to adapt to the norms and rules of the situation (Sproull & Kiesler, 1991). These are challenges of face-to-face communications, and limitations in online interactions. In fact, the absence of social context cues can create conflict in communication because social inhibitions are reduced. Senders may display less sociably desirable behavior and the sender may be less concerned with others.

Symbolic Interactionism Theory

We can also look to sociology to help us make sense of communication interactions. A major perspective in sociology is called symbolic interactionism and it is focused on concrete details of what happens between people in everyday life. We will use symbolic interactionism in Section 3 to help us craft a plan of action for bridging the communication gap between FTF and online communication. For now, we will focus on the theory. Interactionists study how people use and interpret symbols to enhance and improve communication interaction. Further, interactionists use this theory to strive to create a sense of self and to add realism to social situations we experience, working to create and sustain those activities. What does all of this mean? It means we can view social life as a large, complex relationship with countless interactions that allow life to take shape and meaning.

Interactionists create and recreate meanings in a process of interpretation that is prominent in the communicative activity. Individuals communicate using language and symbols to interpret subjective viewpoints, allowing them to make sense of the world. "Symbolic interactionists are often less concerned with objective *structure* than with subjective *meaning* – how repeated, meaningful interactions among individuals come to define the makeup of society." (Carter & Fuller, 2015) The basics of symbolic interactionism are that:

1. People act based on the meanings a particular social and cultural context based on the meanings objects have for them;
2. Interaction occurs within a particular social and cultural context in which physical and social objects (persons), as well as situations, must be defined or categorized based on individual meanings;
3. Meanings emerge from interactions with other individuals and with society; and
4. Meanings are continuously created and recreated through interpreting processes during interaction with others. (Blumer, 1986)

Jan Fernback of Temple University presented a view of interactionism that is perfectly suited for our comparison of FTF and digital conversations. In a 2007 article about a study of symbolic interactionist perspective on online social relations, Fernback suggests an evolving process that begins to address the issue of convenient togetherness without real responsibility (2007). Using interviews and a theoretical consideration of online community, the research effort found that some participants were anxious to discuss additional meaning that would be helpful in online social activity. Networked individualism is a social capital-building approach that focuses interaction on selected individual residents and their immediate social clusters (Wellman, 2012). This peer-to-peer mode of exchange departs from a collectivist approach to community in "normal" mode of online communication, allowing online communicators some of the additional, personal information that is typically available only in FTF activities. Further, the smaller group interactions suggested have a more personal feel and usefulness, striving for fellowship, respect, and tolerance (Fernback, 2007). An important aspect of networked individualism is that online communicators can associate with a church or a job or some other group to transcend physical place, thereby achieving continuity and commitment that is not available in the traditionally large community groups we see in the online environment.

Emotional Intelligence

Beyond the theories mentioned above, our focus begins with the knowledge that there are a variety of definitions of Emotional Intelligence (EI), but most scholars define it as a multifaceted concept that involves a broad range of skills and behaviors. For instance, Goleman (1995) describes EI as a manager's ability to recognize the emotions in self and others and to use this information to make improvements in self-management and relationships with others. Fundamentally, EI is a process of gaining awareness by recognizing personal emotions and those of others. Once emotions are brought to a state of consciousness, EI leads people to use this awareness, or information, skillfully and intelligently in deliberate, purposeful decision-making activities. Berman and West (2008) advance the notion that typically EI skills and behaviors involve the domains of self-awareness, self-management, social awareness, and relationship management (Cherniss, 2001; Druskat, 2001; Hood, 2004). EI is relevant to workplace settings and its influence extends far beyond.

EI's foundation is self-awareness, the ability to recognize your own emotions and to have a strong sense of your own tendencies and abilities. It is important to develop people skills, which typically include active listening, acknowledging others, and mindful speaking (Goleman 2001, 2006). Managers, through EI, can use self-awareness to identify and express their emotions while understanding that being attuned to one's feelings can also be a source of feedback that helps people focus on their strengths, weaknesses, and values, hopefully resulting in self-confidence (Gowing, 2001; Jacobs, 2001; Macaleer, 2002). Berman and West (2008) broadly define self-management as "the ability to accomplish while working through one's emotions and perceptions as they arise and to maintain one's values (such as integrity) and standards in the process." It also includes keeping disruptive emotions in check, thinking clearly under difficult, stressful conditions, having flexibility and adaptability to avoid being rigid in responding to others, and taking tough, unpopular positions on issues (Caudron, 1999; Smigla, 2000; Dearborn, 2002).

EI is the ability to sense, understand, and effectively apply the power and acumen of emotions as a source of human energy, information, trust, creativity and influence (Cooper, 1997). Those who improve their EI can more effectively acknowledge and value feelings in themselves and in others, and can respond to those feelings in an effective way. Paying attention to emotions can save the leader time by directing energies more effectively, and expanding opportunities. EI has three driving forces: build trusting relationships, increase energy and effectiveness, and create the future. Research shows that EI far outweighs IQ and raw brain power as the primary success factor in decision making, creating dynamic organizations, and achieving lifestyle satisfaction and success (Cooper, 1997).

The core capacities of EI are drawn together and put into action in the workplace through the framework of the Four Cornerstone Model of Emotional Intelligence, first envisioned by Ayman Sawaf of Advance Intelligence Technologies (Cooper, 1997). The four cornerstones are: emotional literacy, emotional fitness, emotional depth, and emotional alchemy.

1. Emotional literacy involves developing a clear and useful vocabulary for emotional literacy and recognizing, respecting, and valuing the inherent wisdom of feelings. Emotional honesty, emotional energy, emotional feedback, and practical intuition contribute to emotional literacy.
2. Emotional fitness has trust as one of its key characteristics. Emotional fitness also includes authenticity, resilience, renewal, and constructive discontent. Trust is an actionable emotional strength – something we must feel and act on.
3. Emotional depth has to do with calling forth your core character, identifying and advancing the unique potential and purpose that define your

destiny, manifesting the commitment, drive, initiative, conscience, and accountability, applying integrity, and increasing your influence beyond authority, rank, and title.

4. Emotional alchemy is a blending of forces that enable us to discover creative opportunities and transform lesser ideas into greater ones.

It is important to take a closer look at emotional literacy, which is the ability to recognize, understand, and appropriately express our emotions. Just as verbal literacy is the basic building-block for reading and writing, emotional literacy is the basis for perceiving and communicating emotions. Becoming emotionally literate is akin to learning the alphabet, grammar, and vocabulary of our emotional lives.

The ability to read others – to be aware of their needs and feelings – and to anticipate and address the needs of others, is part of social awareness. It also includes verbalizing emotions experienced by others, acknowledging sensitivity to and consideration for the perspectives of others, and staying in touch with others' feelings. The concept of social awareness is basic to creating environments where diverse people can thrive. A related concept is relationship management, which is the ability to build bonds of alignment, appreciation, and support. Effective relationship management requires communication, teamwork, conflict management skills, and the ability to help people work toward common objectives (Macaleer, 2002).

Self-awareness, social awareness, and relationship management are key management skills, highlighting various behaviors and skills that relevant literature mentions and uses for assessment and measurement. Berman and West (2008) argue that EI acknowledges that although emotions are non-cognitive in nature, people can learn to recognize what they feel. People can develop courses of action that consider a rational dialogue about their own feelings and those of others. Studies show that EI increases with age as a result of learning (Bar-On, 2005). EI workplace training programs show that managers can improve their EI skills (Sala, 2000; Cherniss, 2001), for example, by recognizing a broader range of emotions in themselves and others and by distinguishing sincere smiles from false smiles (Ekman, 2004).

An additional consideration involved with improving EI and using it to the manager's advantage is one of change. For instance, if negative emotions present significant problems and indicate the need for adaptability or change, there are important dynamics to consider – possibly radical change in the organization.

The effect of emotion on radical change dynamics can be best understood by looking at the change process as a number of separate components. There are three critical steps required to achieve radical change: receptivity, mobilization, and learning (Huy, 1999). Receptivity is a person's willingness to consider change and it is characterized as both a state and a process. At any fixed point in time, a person can accept the need for the proposed change if there is an interpretive, attitudinal state on the cognitive and emotional level.

The concrete action a person takes in the direction of change is mobilization. This is the process of rallying and propelling different segments of the organization to undertake joint action and to realize common change goals (Huy, 1999).

Receptivity and mobilization are linked to, or lead to, learning. Individuals learn by thinking and then acting, using the outcome of action to revise their belief systems (Kim, 1993). Receptivity is an observed change, where individuals exhibit various stages of willingness to accept the proposed change, from resigned, passive acceptance to enthusiastic endorsement (Huy, 1999).

Huy's (1999) characterization of the radical change process maintains that when receptivity leads to motivation, individuals and organizations also can learn from the outcomes of the changes they enact, and learning provides a feedback loop from the outcomes of behavioral change back to receptivity. In other words, the learning process is a beginning, but that beginning leads back to using that process to sustain receptivity at the desired level. In turn, sustained receptivity at the correct level leads to continued mobilization and so forth. The key issue, then, is how the radical change process begins and continues.

EI skills of managers in the private sector are covered by most research (2005), while Carmeli (2003) has published one of the few studies examining EI skills among public managers. He found that EI skills augment favorable outcomes and attitudes when sampling chief financial officers in Israel. Matheson's (1999) study of upward mobility in the Australian public service also shows the relevance of EI-related competencies.

Emotional intelligence is prevalent in research of theories of emotion and change. Leaders should focus on emotional considerations that address organizational emotional capability and individual EI. Emotions are relevant to the work of public managers because they are a vital part of the human condition (Goleman, 1995, 1998a; Huy, 1999; Hughes, 2005; Berman, 2008). Leaders need opportunities to find harmony between the feelings of themselves and others, to understand their strengths and weaknesses and how they relate to those of their employees, to come to grips with their own negative emotions and adaptability, and to improve their communication and relationship skills.

Enthusiasm and commitment seem to increase when managers create an atmosphere of understanding and acceptance (Sosik, 1999; Goldsmith, 2004;

Light, 2005). Managers should go beyond focusing on just "people skills," instead fully understanding what these skills are and what they mean in terms of emotional considerations. It is obvious that work practices, policies, and the workplace environment create emotions. A manager's understanding of emotions allows workplace success in terms of norming and shaping motivational interactions with employees, especially when working human resource (HR) management issues or agendas. Berman and West (2008) argued that "acknowledging the place of emotional content in administrative scholarship and practice is not new, but its significance has often been relegated to secondary status, if not ignored." It is time to put EI at the forefront to help managers understand it and use it to their best advantage.

According to Berman and West [citation], there is a new theoretical framework for workplace relationships between managers and those they employ and motivate. The focus is on examining how organizational policies and practices affect manager emotional skills in public organizations. The authors addressed managing EI in U.S. cities. Further, they studied the social skills of public managers and argued the need for better integration of emotions in management theory. Their basis was the range and salience of organizational efforts found in their research.

In their research, Berman and West assessed local governments' perceptions of managerial emotional intelligence, and practices and policies that affect it. They surveyed city managers and chief administrative officers (CAOs) in all 662 U.S. cities with populations over 50,000 (ICMA 2005). The researchers argued that "though few cities offer training programs for managers that are called 'emotional intelligence,' many cities engage in activities that target or address EI skills." This research is insightful for the current examination in this book, because it helps us examine the potential for leaders to improve communication in a variety of areas, such as knowing their own strengths and weaknesses, being more capable of handling negative emotions, increasing their adaptability, and finding ways to improve relationship and communication skills. When leaders communicate, they must assess whether there is a feedback and mentoring process and, if there is, they must determine how the process is related to perceptions of EI. There are also important considerations such as training.

What we now understand is that managers can use adjusted or learned EI skills to help themselves and their organizations adopt strategies and policies that promote efficiency. Organizations have always tried to ensure the emotional skill of managers, but the vehicles have been employee selection, appraisal systems, and formal or informal feedback mechanisms. Managerial EI will not eliminate the need for those avenues of improvement; however, it can provide a mainstream way to transform and improve current practices and enhance existing knowledge. Berman and West's (2008) broad-based framework provides the start of that journey.

Sensemaking Theory

Another theory that is relevant to our study of face-to-face communications is sensemaking. Seven sensemaking properties are the cornerstone of the theory, providing the analytic tools needed to understand the process (Mills & Mills, 2000):

1. **Social Context:** Creating meaning from conversations and relationships.
2. **Personal Identity:** The perception each member has of his or her group role.
3. **Retrospect:** The group interpretation of what has occurred.
4. **Salient Cues:** The group expands cues into explanations.
5. **Ongoing Events:** The group understands the reality that members must act and respond as things continue to change.
6. **Plausibility:** The group makes credible sense (stories) from salient cues and ongoing events within the social context.
7. **Enactment:** Further action is informed by action and learning.

Sensemaking is a social activity that requires decoding messages for ourselves, drawing upon a common language we use and upon the actions we engage in. It is important to share ideas and influence how others make sense of events.

Social context suggests that people delve deeper into relationships than may be possible via the Internet. To that end, the key to success is having a formula that allows you to understand the level at which you're connected and then decide whether that level suits your needs. When seeking this understanding, several things need to be considered:

1. Have you created strong ties?
2. Consider quantity versus how many quality relationships you can reasonably handle in the time you have available from day to day.
3. Employ people sensemaking (DiMicco & Millen, 2008) to narrow and focus your networking efforts for the greatest effect, whatever that is. In other words, what is your goal: counting connections or achieving quality connections?

4. Search for like-mindedness, which often gets lost in the shuffle in the large number of impersonal Internet communications.

Is your relationship characterized by strong or weak ties? Weak ties promote innovation and strong ties promote performance. Several characteristics distinguish strong from weak ties when analyzing relationships. In essence, as the characteristic increases, ties get stronger. Consider age of the relationship, frequency of contact, emotional attachment, reciprocity, and kinship.

Weak ties are characterized by absent or infrequent contact and lack of emotional closeness and reciprocal services. They are considered best suited for innovation (Granovetter, 1973). Strong ties are those that demonstrate strong investment of time and reciprocity. Barry Wellman (1997) argued that sets of actors who maintain strong ties are more likely to trust each other in knowledge sharing, behavior modeling, and in the decision-making process (Wellman & Wortley, 1990). The research suggests that organizations should create as many weak ties as possible to foster open communications and innovation. However, to influence behavior and affect performance, the organization should look to create strong ties.

In reality, people only have time and attention to handle a few strong ties. Determine which are the most important and cultivate them. A small number of strong ties may serve you better than hundreds of weak ties.

Sensemaking is about gaining a general understanding of who a person is (DiMicco & Millen, 2008). Gathering information face-to-face makes it possible to fill in the gaps that are left when people filter what they share on the Internet. That information leads to a coherent picture of a person's identity and allows connected people to build trust and reputation so vital to strengthen ties.

Like-mindedness is important to relationships that endure. People tend to get a false sense of security from the enormous number of contacts we generate through Internet-based networking. This false security can make it easy to neglect the talking time necessary to get to know people as individuals (Curtis & Lewis, 2010). By identifying aspects of like-mindedness, quality relationships can be built on real connections where there is commonality between people.

Communication Accommodation Theory (CAT)

Now we move beyond EI and sensemaking to the fundamental theory for this body of work, CAT. Any successful interaction requires that we adjust our communication approach to accommodate others and to allow flexibility in understanding and information exchange. We communicate differently with different audiences in different settings. Our approach, and our openness, differs for conversations that involve our parents, our children, our significant others, our bosses, and our acquaintances. We make adjustments for communicative or behavioral norms to avoid being misunderstood or to guard against being seen as rude, offensive, or insensitive. This is important because failed communicative adjustments lead to problematic, adversarial or dissatisfying experiences.

The best way to find success in making communication adjustments is through the use of CAT. CAT energizes intergroup dynamics by dealing with various ways personal and social identity emerges in social encounters, interactions, and even negotiations (Giles, 2008). The theory states that people adjust their speech, vocal patterns, and gestures to accommodate others when they communicate (Gallois & Giles, 2015). CAT is a way to examine why people work to emphasize or minimize the differences between themselves and others during non-verbal and verbal communication. The theory also focuses on important interaction considerations like language, context, identity, and intergroup and interpersonal factors.

CAT focuses on convergence and divergence in communication. Convergence is what we call the strategies that individuals use to adapt to each other's communicative behaviors, while divergence is about individuals stressing and managing the non-verbal and speech differences between themselves and others.

There are various ways to put CAT into practice. For instance, when giving a speech, we can adjust how we present the information to satisfy the particular audience's collective comprehension ability. Changing the rate of speech, changing the terms or phrases we use, adjusting the volume, changing the tempo, and using repetition are all ways in which we can enhance the interaction.

Several CAT strategies are useful for the discovery and subsequent discussions in this book. By employing interpersonal control, interpretability, discourse management, and emotional expression as our strategies, we can enhance our communication abilities. Interpersonal control is about how much the sender in a communication controls the receiver or, in a more general sense, how one person in an interaction controls the other. This is important because sender and receiver roles change during communication activities. Interpersonal control, then, is about managing or regulating another's thoughts, feelings, or actions (Stets, 1991). Interpretability is a strategy that individuals use with the intent to make their communication more understandable to the receiver (Jones, Woodhouse et al. 2007). This strategy is important because we should always work to make sure that our partners in a communication can understand or interpret what we are delivering to them. It's important to use visual cues, emotions, and other characteristics of the receiver to ensure that our message is delivered.

Discourse management is the ability to organize the information we are sending or delivering, and to repair any difficulties or breaks in the communication. This is the ability to move the conversation or communication in a specific direction to arrive at a successful information exchange. Finally, emotional expression is just what it sounds like: how people express, regulate, experience, and influence emotions. Emotional expression is about understanding and managing our emotions and the emotions of others during a communicative interaction. Emotional expressions are observable and they take the form of verbal and non-verbal behaviors. That means we are talking about audible sounds and facial movements in addition to obvious emotional reactions such a laughing, crying, smiling, or scowling.

Table 1 examines CAT strategies as they relate to receiver activity levels. The information is helpful in identifying effective and ineffective communication activities.

Table 1. Adapted from "Coding System for Communication Strategies," Effective nurse parent communication: A study of parents' perceptions in the NICU environment

Sender-Receiver Relationships			
	Receiver Activity Level: How much does the receiver engage with the message and its sender?		
Strategy	**High**	**Moderate**	**Low**
Interpersonal Control	• Emphasis on personal power or position • Too formal • Lack of connection to or acceptance of the message	• Genuinely interested in the message • Participates in feedback • Polite • Respectful • Encouraging	• In awe of the sender • Failure to share information or feedback • Surrenders too much authority to the sender
Interpretability	• Vague or ambiguous responses • Confusing or inaccurate information • No explanations	• Clear, direct, honest • Straightforward and to the point • Plenty of explanations • Verify understanding of the message	• Topic is too simple to hold your interest • Being "talked down to" • Feeling of inferiority
Discourse Management	• Dominate the conversation • Fails to listen • Does not let others speak	• Ask questions • Ask opinions • Open to new ideas and topics • Listen first, talk second • "Chat" relationship	• Responds only when asked • Allows sender to determine what is discussed

continued on following page

Table 1. Continued

Sender-Receiver Relationships			
	Receiver Activity Level: How much does the receiver engage with the message and its sender?		
Strategy	**High**	**Moderate**	**Low**
Emotional Expression	• Hostile • Unsupportive	• Show empathy • Establish caring interaction • Give reassurances about the message	• Too much sympathy • Lack of real information exchange

(Jones, Woodhouse et al. 2007)

This table is adapted from a study published in 2007 that examined effective nurse to parent communication, finding that effective communication was accommodative and more interpersonal while ineffective communication was generally under-accommodative and more intergroup (Jones, Woodhouse et al. 2007). What is useful for the analysis we are undertaking is this study's evaluation and findings where communication strategies are concerned. When we used CAT strategies, the analysis was about perceptions of effective and ineffective communication. The findings suggest that effective communication was derived from shared management of the interaction and appropriate support and reassurance by the nurses. Ineffective communication was characterized by under-accommodation and more intergroup exchanges.

Study participants believed that the communication was more effective when the interaction was more equal (Jones, Woodhouse et al. 2007). Participants also mentioned discourse management and emotional expression most often when highlighting successful communications and valued communication that was two-way and involved informal chatting. Interpretability was commonly mentioned by participants as contributing to ineffective communication, characterized by information that was inconsistent, vague, conflicting, indirect, or confusing. This is important for the current evaluation, because these things can be discovered and possibly corrected during face-to-face communication. By contrast, they are frequently unknown in digital conversations.

The table as adapted for this book focuses on CAT strategies of interpersonal control, interpretability, discourse management, and emotional expression from the nurse communication study. The new table evaluates sender-receiver relationships, characterizing receiver activity levels in terms of each strategy. This is a view of communication as a balancing act, where high or low activity levels may not bring the optimum communication, and moderate activity gives us the most reward for our efforts. Based on each strategy, the table attempts to explain how and why receiver activity levels are high, moderate, or low. This is important because the table suggests that moderate levels of receiver activity are more accepting of the message, and that could possibly lead to greater understanding. The best outcome would be increased feedback because the message is clear and understandable.

The information in the table demonstrates the importance of emotions in communications, which relates to our earlier discussion of EI. This adaptation of CAT and EI energizes our journey to understand and address gaps in communication between face-to-face and online interactions.

The preceding examination of theory, especially EI, sensemaking, and CAT, solidifies the foundation upon which we view communication activities. The high-touch approach we seek is about gaining awareness through EI and its concepts, and using sensemaking theory to create and nurture social activities characterized by a common language and shared ideas. CAT concepts can assist communicators to minimize differences that could hinder interactions, such as language, context, identity, and other factors.

REFERENCES

Akdere, M., & Roberts, P. B. (2008). Economics of Social Capital: Implications for Organizational Performance. *Advances in Developing Human Resources*, *10*(6), 802–816. doi:10.1177/1523422308325007

Argyle, M., & Dean, J. (1965). Eye-contact, distance and affiliation. *Sociometry*, *28*(3), 289–304. doi:10.2307/2786027 PMID:14341239

Bandura, A. (1986). *Social foundations of thought and action: a social cognitive theory*. Englewood Cliffs, NJ: Prentice-Hall.

Bandura, A. (1989). Human agency in social cognitive theory. *The American Psychologist*, *44*(9), 1175–1184. doi:10.1037/0003-066X.44.9.1175 PMID:2782727

Bar-On, R. (2005). The Bar-On Model of Emotional-Social Intelligence. *Psichothema, 17*.

Bearman, T. C., Guynup, & Milevski. (1985). Information and Productivity. *Journal of the American Society for Information Science, 36*(6), 369.

Berman, E. M., & West, J. P. (2008). Managing Emotional Intelligence in U.S. Cities: A Study of Social Skills among Public Managers. *Public Administration Review*, *68*(4), 742–758. doi:10.1111/j.1540-6210.2008.00912.x

Blumer, H. (1986). *Symbolic interactionism: Perspective and method.* Univ of California Press.

Brown, M. A. Sr, & Schario, T. A. (2014). *Social Media 4EVR: Identifying, Achieving, & Nurturing Social Capital.* Yorktown, VA: CreateSpace Independent Publishing Platform.

Carmeli, A. (2003). The relationship between emotional intelligence and work attitudes, behavior and outcomes: An examination among senior managers. *Journal of Managerial Psychology, 18*(8), 788–813. doi:10.1108/02683940310511881

Carter, M. J., & Fuller, C. (2015). *Symbolic interactionism.* Academic Press.

Caudron, S. (1999). The Hard Case for Soft Skills. *Workforce, 78*(7), 60–66.

Cerulo, K. A. (1990). To Err Is Social: Network Prominence and Its Effects on Self-Estimation. *Sociological Forum, 5*(4), 619–634. doi:10.1007/BF01115394

Chatman, E. A., & Pendleton, V. E. (1995). Knowledge gap, information-seeking and the poor. *The Reference Librarian, 23*(49-50), 135–145. doi:10.1300/J120v23n49_10

Cherniss, C., & Goleman, D. (2001). *The Emotionally Intelligence Workplace: How to Select for, Measure, and Improve Emotional Intelligence in Individuals, Groups, and Organizations.* San Francisco: Jossey-Bass.

Chung, K. S. K., Hossain, L., & Davis, J. (2007). Individual performance in knowledge intensive work through social networks.*Proceedings of the 2007 ACM SIGMIS CPR conference on Computer personnel research: The global information technology workforce.* St. Louis, MO:ACM. doi:10.1145/1235000.1235036

Cooper, J., & Carlsmith, K. M. (2001). *Cognitive Dissonance A2. In International Encyclopedia of the Social & Behavioral Sciences* (pp. 2112–2114). Oxford, UK: Pergamon.

Cooper, R. K. (1997). Applying Emotional Intelligence in the Workplace. *Training & Development, 51*(12), 31–38.

Curtis, G. S., & Lewis, G. (2010). *Well connected: An unconventional approach to building genuine, effective business relationships.* San Francisco, CA: Jossey-Bass.

Davis, T. R. V., & Luthans, F. (1980). A Social Learning Approach to Organizational Behavior. *Academy of Management Review*, *5*(2), 281–290.

Dearborn, K. (2002). Studies in Emotional Intelligence Redefine Our Approach to Leadership Development. *Public Personnel Management*, *31*(4), 523–530. doi:10.1177/009102600203100408

DiMicco, J. M., & Millen, D. R. (2008). People Sensemaking with Social Networking Sites. *Sensemaking Workshop, CHI 2008*. Florence, Italy: IBM Research.

Druskat, U., & Wolff, S. (2001). Building the Emotional Intelligence of Groups. *Harvard Business Review*, *79*(3), 81–90. PMID:11246926

Eastin, M. S., & LaRose, R. (2000). Internet Self-Efficacy and the Psychology of the Digital Divide. *Journal of Computer-Mediated Communication*, *6*(1).

Ekman, P. (2004). *Emotions Revealed: Recognizing Faces and Feelings to Improve Communication and Emotional Life*. New York: Henry Holt.

Erickson, C. L., & Jacoby, S. M. (2003). The effects of employer networks on workplace innovation and training. *Industrial & Labor Relations Review*, *56*(2), 203–223. doi:10.1177/001979390305600201

Evan, M., & Berman, J. P. W. (2008). Managing Emotional Intelligence in U.S. Cities: A Study of Social Skills among Public Managers. *Public Administration Review*, *68*(4), 742–758. doi:10.1111/j.1540-6210.2008.00912.x

Fernback, J. (2007). Beyond the diluted community concept: A symbolic interactionist perspective on online social relations. *New Media & Society*, *9*(1), 49–69. doi:10.1177/1461444807072417

Field, J. (2003). *Social Capital*. London: Routledge.

Fowler, S. W., Lawrence, T. B., & Morse, E. A. (2004). Virtually Embedded Ties. *Journal of Management*, *30*(5), 647–666. doi:10.1016/j.jm.2004.02.005

Gallois, C., & Giles, H. (2015). *Communication Accommodation Theory. The International Encyclopedia of Language and Social Interaction*. John Wiley & Sons, Inc.

Giles, H. (2008). *Communication accommodation theory*. Sage Publications, Inc.

Goldsmith, S., & Eggers, W. D. (2004). *Governing by Network: The New Shape of the Public Sector*. Washington, DC: Brookings Institute Press.

Goldsmith, S., & Eggers, W. D. (2004). *Governing by network: the new shape of the public sector*. Washington, DC: Brookings Institution Press.

Goleman, D. (1995). *Emotional Intelligence*. New York: Bantam.

Goleman, D. (1998a). *Working with Emotional Intelligence*. New York: Bantam.

Goleman, D. (Ed.). (2001). *The Emotionally Intelligent Workplace. An EI-Based Theory of Performance*. San Francisco: Jossey-Bass.

Goleman, D. (2006). *Social Intelligence*. New York: Bantam.

Gowing, M. K. (Ed.). (2001). *The Emotionally Intelligent Workplace. Measurement of Individual Emotional Competence*. San Francisco: Jossey-Bass.

Granovetter, M. S. (1973). The Strength of Weak Ties. *American Journal of Sociology*, *78*(6), 1360–1380. doi:10.1086/225469

Gunawardena, C. N. (1995). Social presence theory and implications for interaction and collaborative learning in computer conferences. *International Journal of Educational Telecommunications*, *1*(2), 147–166.

Harmon-Jones, E. (2012). *Cognitive Dissonance Theory*. Academic Press.

Hatala, J.-P. (2006). Social Network Analysis in Human Resource Development: A New Methodology. *Human Resource Development Review*, *5*(1), 45–71. doi:10.1177/1534484305284318

Hatala, J.-P., & Fleming, P. R. (2007). Making Transfer Climate Visible: Utilizing Social Network Analysis to Facilitate the Transfer of Training. *Human Resource Development Review*, *6*(1), 33–63. doi:10.1177/1534484306297116

Hatala, J.-P., & Lutta, J. G. (2009). Managing Information Sharing Within an Organizational Setting: A Social Network Perspective. *Performance Improvement Quarterly*, *21*(4), 5–33. doi:10.1002/piq.20036

Hood, C., & Lodge, M. (2004). Competency, Bureaucracy, and Public Management Reform: A Competency Analysis. *Governance: An International Journal of Policy, Administration and Institutions*, *17*(3), 313–333. doi:10.1111/j.0952-1895.2004.00248.x

Hughes, J. (2005). Bringing Emotion to Work: Emotional Intelligence, Employee Resistance and the Reinvention of Character. *Work, Employment and Society*, *19*(3), 603–625. doi:10.1177/0950017005055675

Huy, Q. N. (1999). Emotional Capability, Emotional Intelligence, and Radical Change. *Academy of Management Review*, *24*(2), 325–345.

Jacobs, R. L. (Ed.). (2001). *The Emotionally Intelligent Workplace. Using Human Resource Functions to Enhance Emotional Intelligence*. San Francisco: Jossey-Bass.

Jones, L., Woodhouse, D., & Rowe, J. (2007). Effective nurse parent communication: A study of parents perceptions in the NICU environment. *Patient Education and Counseling*, *69*(1), 206–212. doi:10.1016/j.pec.2007.08.014 PMID:17936549

Kim, D. H. (1993). The link between individual learning and organizational learning. *Sloan Management Review*, *35*(Fall), 379–500.

Light, P. C. (2005). *The Four Pillars of High Performance: How Robust Organizations Achieve Extraordinary Results*. New York: McGraw-Hill.

Linking Emotional Intelligence and Performance of Work. (2005Mahwah, NJ: Lawrence Erlbaum.

Lowenthal, P. R. (2009). The Evolution and Influence of Social Presence Theory on Online Learning. In *Social Computing: Concepts, Methodologies, Tools, and Applications*. Academic Press.

Ma, L. (2007). *"I'll trust you if I expect you to trust me": An analysis of interpersonal trust, friends, and social interactions within social networks*. Tufts University.

Macaleer, W. D., & Shannon, J. B. (2002). Emotional Intelligence: How Does it Affect Leadership? *Employment Relations Today*, *29*(3), 9–19. doi:10.1002/ert.10047

Matheson, C. (1999). The Sources of Upward Mobility within Public Sector Organizations. *Administration & Society*, *31*(4), 495–524. doi:10.1177/00953999922019210

Mew, L. (2009). *Online Social Networking: A task-person-technology fit perspective*. The George Washington University.

Mills, J. H., & Mills, A. J. (2000). *Sensemaking and the Gendering of Organizational Culture*. Montreal, Canada: ASAC-IFSAM Conference.

Newell, S., & Swan, J. (2000). Trust and inter-organizational networking. *Human Relations*, *53*(10), 1287–1328.

Nyhan, R. C. (2000). Changing the Paradigm: Trust and its Role in Public Sector Organizations. *American Review of Public Administration*, *30*(1), 87–109. doi:10.1177/02750740022064560

Otis, B. (2007). *Factors in Social Computing Related to Worker Productivity. Master of Science Capstone Report*. University of Oregon.

Palo Alto Networks. (2009). *The Application Usage and Risk Report: An Analysis of End User Application Trends in the Enterprise*. Sunnyvale, CA: Palo Alto Networks.

Perry, R. (2006). Diffusion theories. In Encyclopedia of Sociology. New York: Macmillan Reference.

Putnam, R. D. (1993). *Bowling Alone*. New York: Simon & Schuster Paperbacks.

Rhodes, J., Lok, P., Hung, R. Y.-Y., & Fang, S.-C. (2008). An integrative model of organizational learning and social capital on effective knowledge transfer and perceived organizational performance. *Journal of Workplace Learning*, *20*(4), 245–258. doi:10.1108/13665620810871105

Sabatini, F. (2009). Social capital as social networks: A new framework for measurement and an empirical analysis of its determinants and consequences. *Journal of Socio-Economics*, *38*(3), 429–442. doi:10.1016/j.socec.2008.06.001

Sala, F. (2000). *Do Programs Designed to Increase Emotional Intelligence at Work - Work?*. Academic Press.

Segrest, S. L., Domke-Damonte, D. J., Miles, A. K., & Anthony, W. P. (1998). Following the crowd: Social influence and technology usage. *Journal of Organizational Change Management*, *11*(5), 425–445. doi:10.1108/09534819810234841

Shetzer, L. (1993). A social information processing model of employee participation. *Organization Science*, *4*(2), 252–268. doi:10.1287/orsc.4.2.252

Short, J., Williams, E., & Christie, B. (1976). *The social psychology of telecommunications*. New York: Wiley.

Smigla, J. E., & Pastoria, G. (2000). Emotional Intelligence: Some Have It, Others Can Learn. *The CPA Journal*, *70*(6), 60–61.

Sosik, J. J., & Megerian, L. E. (1999). Understanding Leader Emotional Intelligence and Performance. *Group & Organization Management*, *24*(3), 367–390. doi:10.1177/1059601199243006

Sproull, L., & Kiesler, S. (1991). *Connections: New ways of working in the networked organization*. Cambridge, MA: MIT Press.

Stets, J. E. (1991). Psychological aggression in dating relationships: The role of interpersonal control. *Journal of Family Violence*, *6*(1), 97–114. doi:10.1007/BF00978528

Weiner, B. (1994). Integrating Social and Personal Theories of Achievement Striving. *Review of Educational Research*, *64*(4), 557–573. doi:10.3102/00346543064004557

Wellman, B. (1997). *An electronic group is virtually a social network. In Culture of the Internet* (pp. 179–205). Hillsdale, NJ: Lawrence Erlbaum.

Wellman, B. (2012). Networked individualism: how the personalized internet, ubiquitous connectivity, and the turn to social networks can affect learning analytics.*Proceedings of the 2nd International Conference on Learning Analytics and Knowledge*. ACM. doi:10.1145/2330601.2330603

Wellman, B., & Wortley, S. (1990). Different Strokes from Different Folks: Community Ties and Social Support. *American Journal of Sociology*, *96*(3), 558–588. doi:10.1086/229572

Williams, D. (2006). On and Off the 'Net: Scales for Social Capital in an Online Era. *Journal of Computer-Mediated Communication*, *11*(2), 593–628. doi:10.1111/j.1083-6101.2006.00029.x

Zack, M. H., & McKenney, J. L. (1995). Social context and interaction in ongoing computer-supported management groups. *Organization Science*, *6*(4), 394–422. doi:10.1287/orsc.6.4.394

Chapter 4
SNIP:
High Touch Approach to Communication

OVERVIEW

This chapter uses a high-touch approach to social capital, focusing on a new and improved way to affect social networking participation and success. Social Networking and Individual Performance, or SNIP, was created to evaluate and achieve success in digital communication with an emphasis on reinforcing organizational training and improving internal and external communications through social media. We will examine the relationship between social networking and individual performance, paying attention to key considerations like individual perceptions of productivity, levels of trust, and the ability to conduct bonding, bridging, and linking activities. The chapter will move through a focus on energizing the use of social capital to comparing social networking and social media. We know that social networking builds trust and seeks engagement, while social media provides tools to facilitate the communication. The chapter concludes with the history and value of SNIP, a social networking participation model that can provide information, techniques, and examples that address real or perceived benefits in order to assist in the social networking journey. This action plan for success provides a tool that can build and strengthen social capital for people and organizations.

DOI: 10.4018/978-1-5225-1897-6.ch004

Social Networking and Individual Performance

The high-touch communication approach contained herein is about developing social capital. Examinations of social capital in the online era have been dominated by researchers who have imported measurements from different media, often in functions other than social networking or communication (Williams, 2006). This author's efforts have focused on finding a new way to measure performance in social networking, and the benefits of those measurements may translate to FTF communication. Social Networking and Individual Performance, or SNIP, is a concept the author created to evaluate and achieve success in digital communication. SNIP, further described in Figure 1, is about reinforcing organizational training and improving internal and external communications through social media.

Success also relies on energizing social capital and taking advantage of the networks and relationships that are nurtured through exchanged value. A comparison of social networking and social media creates a baseline for effectively managing online activities. SNIP offers an opportunity for organizations to take advantage of unfettered, two-way communication with important audiences. The beauty of social networking lies not only in its immediacy, but in its ability to get and give feedback in real time.

Figure 1. A note on social construction of reality

A Note on Social Construction of Reality

SNIP has an educational value in that it is primarily focused on public administration environments; however, the social networking experiences are relevant for any number of industries. I hope that one day there will be a scored instrument to provide a social media aptitude measure and offer a vehicle by which social networking participation can be analyzed and improved. Understanding the relationship between social networking and individual performance and employing a social networking plan are key steps to finding success in communication activities.

Dr. Michael A. Brown Sr.

The social media revolution is upon us and it's not slowing down. Organizations, whether they want to or not, are going social on the inside, so it's important to listen to the digital conversation at every opportunity. Listening to the conversation strategically helps to engage stakeholders and friends using the channels they prefer.

Social networking and social media are great tools that enable organizations to unleash communication and improve listening ability. Social capital is built this way and engagements start to take on organizational importance. This book is about accepting those opportunities.

Relationship between Social Networking and Individual Performance

Organizations that want to be successful in social networking activities should understand that the keys to participation lie in individual behavior. Key considerations are individual perceptions of productivity, levels of trust, and the ability to conduct bonding, bridging, and linking activities.

Social networking is a voluntary process in which people are faced with choices. Organizations cannot achieve full success in social networking, however, unless they address acceptance and rejection of participation. Why? Because being social requires the exchange of valuable information and the resulting two-way interaction between participants. The relevant questions for organizations have to do with finding out how people determine their connections and how they decide on how much time and effort they are willing to commit to the activity.

One basis for stressing this social networking and individual performance relationship is in the work of Svendsen and Sorenson (2006), which examined whether the levels of individual perception of productivity and levels of trust would be influenced by measuring social capital as the density of voluntary networking activities. The focus of social networking activities should be on the capacity of individuals to perform the necessary tasks. SNIP offers a framework organizations can use to affect creativity, idea exchange, and communication effectiveness.

It is important to note that a few key theories relate directly to this relationship between networking and performance and getting "social." Social networking characteristics can be examined through the Hawthorne Effect, social learning theory, and social network analysis (SNA). A primary focus is placed on behavioral intention which is defined as the extent to which a person believes that the technology is capable of being used advantageously and that the technology provides positive expected outcomes. Prior research related to these theories provides a participation foundation, one built on the understanding that social networking and individual performance share the activities of bonding (establishing strong ties), bridging (establishing informal ties), and linking (establishing voluntary ties).

In networking, strong ties are clusters of people in strong relationships who know what one another know (Granovetter 1973, 1982). These typically long-term ties rely on collaboration, confidence, and trust between people and networks. Weak ties are characterized by having no strong or established social group. The knowledge creation process in the network is supported by strong ties because they deliver strategic capabilities based on their personal connections, and they assist with finding new and valuable information opportunities. The benefit of weak ties is that they introduce new information into the network.

Organizations can analyze employee behavioral intention to focus on strong ties that encourage the desired behaviors for maximum effectiveness. Organizational keys to improving social networking activities rely on three things:

1. Concrete incentives to participate in the kind of social media the organization values;
2. Improved definition of the work required through task analyzability; and
3. Continuous feedback through structured reinforcement.

The Importance of Energizing Social Capital

As social networking activities are improved, energizing social capital means finding and taking advantage of meaningful engagements. Organizations that want a good corporate social media focus should be engaged in controlling and expanding their brand by relaying messages quickly while the subject, and the public's attention, is prevalent in the community conversation. It is important to generate conversation and gather feedback to build relationships and foster information sharing among employees.

The benefits begin with increased traffic and interaction on the organization's website or social media platform. Effective two-way communications can help organizations develop an understanding of their target audiences. That understanding is important to finding new and improved ways to find, collect, and share information. Social networking can develop an ability to reach audiences that are unavailable by other means, increasing the organization's reach in the community of interest. Finally, social networking develops transparency with important internal and external publics.

Bridging and bonding are types of social capital according to Robert Putnam (Putnam, 1993). Putnam says that bonding occurs when people socialize with others who have something in common with them: same age, race, religion, etc. Bonding is important, but Putnam contends that bridging is needed to create interacting societies in a diverse multi-ethnic country. Bridging is making friend with people who don't have something in common with us; they are not like us, they don't like our favorites, they don't eat what we eat, and they don't drive what we drive. However, both types of social capital are continuously at play, strengthening each other as online networks are created.

Social capital grows from relationships between and among organizations, individuals, communities, or societies. Social capital building can create close interpersonal relationships because of the valuable resources that are involved. Social networking relies on social capital and has some of the same characteristics; however, there may be no face-to-face communication and the relationships are more informal than those typically created in social capital building.

Comparing Social Networking and Social Media

Social networking is about trust and engagement. People form networks to build trusting relationships that are geared to reduce risk in business or personal decisions by taking advantage of the skills and abilities of others. In terms of engagement, they want to share experiences and seek advice that they hope will make them better decision makers. Social networking is about the "why" of social relationships.

Social media provides the tools and venues that are used to communicate socially. These platforms transcend the boundaries of time or location. Using these platforms, organizations and people can send or receive information effortlessly, connecting with anyone participating in social media anywhere in the world.

But social networking can be so much more for organizations. Consider the individual performance implications. Important discoveries can be made by changing the lens by which we view social networking. SNIP takes a different approach from what is currently evident in the audiences/industries discussed in this proposal. This suggests an approach significantly different from the norm in social networking and social media circles today. It allows the destination, not the path or platform, to be the focus of social media participation. Organizations should look beyond near-term financial benefits and leverage short-term costs (either time or expenses) that promote positive social and environmental change.

SNIP addresses the challenges of understanding social networking and social media platforms enough to assess the relevance of various approaches. A broader vision of social media participation and performance implications is suggested, using a goal-setting focus and social media plan advocated in the SNIP methodology. SNIP attempts to move the traditional frame of reference from solely which platform to use – Facebook, LinkedIn, Twitter, etc. – to a view of goal achievement and audience targeting.

Social Networking Discovery Opportunities

SNIP provides a social networking participation model that can help organizations predict and understand the value proposition that affects participation. Innovation adoption, governing by network, and social capital are important theories in developing an understanding of social networking behavior. Performance may be influenced when people are presented with evidence of a return on the investment of their time and effort.

SNIP provides information, techniques and examples that address real or perceived benefits in order to assist in the social networking journey. This book is a new view of social networking that establishes and articulates the "why" of participation to help organizations find success. The digital conversation is uncontrolled and largely unorganized, an organic, complex series of communications in open space. Social networking uses that dynamic entity to highlight ways to connect and share content with anyone, anywhere.

These communications are simply based on opportunity. When people are presented with evidence of a return on the investment of their time and effort, participation becomes attractive. People may be inclined to participate even if there is only the hint of benefit. This value-based approach is what make social networking relevant and what allows organizations to seek and encourage social media behaviors. This approach is about getting in touch with social networking and social media thought processes.

SNIP offers an action plan for success. The historical perspective and theoretical evolution offered herein shows the growth of maturity of social media. That is followed by information about tools and techniques that build social capital and engage audiences in a meaningful way.

This is a call to action. There is an exciting digital conversation going on, and there are various ways to engage in social networking for success. Please join us in this exciting endeavor and let's communicate anytime, anywhere.

Research is underway for further development of SNIP. Once realized, organizations, researchers, and academicians will have a new source for social networking methodology through SNIP. A proposed scored survey instrument, research findings, relevant discovery, and discussion brings social networking into focus. The current body of work focuses on the end state of social networking activities rather than the social media platform, an important distinction. SNIP does not focus on "how to" use platforms, instead it addresses deciding on an end state and devising a strategic approach to arrive at that destination. There are few references that focus on communication and social networking in the way that I do through SNIP and EVR.

Social capital theory suggests that the efficiency of society can be improved by facilitating coordinated actions (Putnam, 1993). Social networking provides a vital source of information and numerous opportunities for all participants to build social capital. Social media tools are increasingly provided by organizations to improve business processes, create new business, and enhance the lives of employees. The new and innovative ways to stay in touch with the environment and peers through social networking makes for dynamic communication with internal and external audiences. Organizations that commit to social networking also commit to allowing employees to spend an unspecified number of hours making connections and joining communities. Many companies have embraced the fact that there are available benefits in terms of improved communication and morale, and in terms of connecting people on related projects or with related responsibilities (Madden & Jones, 2008). The Pew Internet & American Life Project (2008) found that "Wired and Ready Workers" have improved their work lives through information and communications technology (ICT):

- 80% say these technologies have improved their ability to do their job.
- 73% say these technologies have improved their ability to share ideas with co-workers.
- 58% say these tools have allowed them more flexibility in the hours they work.

The Wired and Ready Workers also note negative impacts of ICTs in the study:

- 46% say ICTs increase demands that they work more hours.
- 49% say ICTs increase the level of stress in their job.
- 49% say ICTs make it harder for them to disconnect from their work when they are at home and on the weekends.

The Pew Internet report establishes that 96% of employed adults are in some way making use of new communications technologies, either by going online, using e-mail, or owning a cell phone (Madden & Jones, 2008). Organizations need to find ways to encourage participation on a higher level, address any issues with acceptance of social networking policies, and address any negative perceptions that stand in the way of people "getting social" to their fullest capabilities.

How can organizational leadership address social networking participation? The foundation provided through relevant theories in this book can help to define ways to create a decision matrix, which could lead to an active, lasting participation in an organization's chosen social networking activity. The decision matrix would provide a framework identifying determinants and opportunities that affect participation. As social capital builds in an effective social networking environment, the density of social networking activity should increase.

Social Networking vs. Social Media

The distinction between social networking and social media must be made clear before introducing a new focus. Social media refers to communication interactions which create, share, and exchange information and ideas in virtual communities and networks. If regular media is a one-way street, then social media is a two-way street that delivers unlimited ability to comment and discuss and offers a fast way to read the newspaper or digest a TV report.

Every day, people reach out to others so that we can reduce risk in personal and business decisions by taking advantage of their expertise. Social media is about using different tools, or platforms, to communicate socially. The tools allow us to communicate and store information without limitations of location or time. Anyone can connect with anyone anywhere if both parties are participating in social networking or social media.

This new view of social networking suggests identifying, achieving, and nurturing social capital. Organizations now have a partner in improving the pursuit of social networking by using social capital to leverage value propositions in two-way, online communications. There is benefit in the pursuit of shared expectations that focus on finding common value and a worthy return on investment of each party's contribution of time and effort.

The social network focus can be made clearer with the existence of connectedness and dependency. Connectedness is about being in touch with people who are similar in thought or action. In this way, connectedness is a value that powers a challenging adventure into digital conversations. Dependency is valuable, but it is sometimes viewed in a negative light. That's because some believe that people are dependent on posting or tweeting the most insignificant, most negative, and craziest bits of information imaginable. That can be true, but it can also be true that dependency is as positive as connectedness.

Why? Social media is a way to get in touch anytime, anywhere with organizations and people who have similarities and ask questions, share information, and take the risk of getting to know each other. That is exciting and invigorating at times, and that is a reason that some can't resist sharing anything, or everything. Think about Maslow's Hierarchy of Needs, in which love, belonging, and esteem are primary concerns. People need friendship, family, confidence, and respect to feel good about themselves. Social media can be an outlet for those feelings.

Whatever participation choice is made should be based on an understanding of both benefits and risks. Start by weighing the pros and cons of social media to understand negative AND positive implications. Organizations should keep in mind that the digital conversation is never ending and it may be harder to catch up later if they decide not to participate. That advice is good for individuals as well.

Social networking is already being used by many organizations because they believe that the benefits of increased or improved communications outweigh the potential problems associated with social networking. Organizations want employees to exhibit behaviors that support social networking efforts. That means there can be performance implications. To address these issues, organizations should base social networking activities on task analyzability, modeling principles, and structured reinforcement.

SNIP can provide additional tools that can support business and employee development for organizations involved in social networking. If employees perceive that there is some return on the investment of their time or attention in terms of improved skills or the availability of new challenges or increased standing in the firm, they may be more inclined to participate in the organization's social media of choice. Many organizations want to understand the social networking return on investment. An improved understanding of employee perceptions and motivations for participation provides a better chance of arriving at social networking that permeates the culture. Igbaria and Tan (1997) demonstrated the impact of social networking acceptance on employer and employee perceptions of performance and productivity. Some of that impact will be investigated here.

Social Networking

There is a relationship between social networking and perception of individual performance (Brown, 2011). Organizations seeking improvements in social networking activities must first understand the nature of the network. This understanding is vital because improvement requires identifying levels of participation, setting a standard or level for desired participation, and convincing employees that their participation has the potential to improve their job performance. Also, organizations should ensure their social networking activities are efficient. Efficiency is best achieved by providing the highest quality task analyzability, using modeling principles to affect behavior, and employing structured reinforcement to provide continuous feedback.

In the search for improvement and efficiency, SNIP focuses on five types of relationships.

1. Relationship between trust in organizational leadership and participation in social networking.
2. The impact of perceived improvement potential (PIP) on participation in social networking.
3. The relationship between modeling opportunities and social networking participation.
4. The importance of task analyzability.
5. Structured reinforcement as it relates to individual improvement.

Social networking activities should focus on whether individuals see a value proposition in terms of their performance. In other words, do individuals believe that they will improve their skills, increase their promotion chances, or get some other benefits as a result of their participation? Will that value proposition convince them to increase their participation? These questions reflect the notion of individual impact (Igbaria & Tan, 1997), which refers to the influence of online interactions on the perceived performance of the individual on the quality of his or her decision making.

As organizations assess their social networking programs and their outcomes, it is important to understand the role of trust in leadership and of perceptions of individual improvement. Any investigation should attempt to address whether social media engagement and networking foster a supportive and cooperative structure and result in increases in generalized trust. The company should seek a trusting structure that will lead to an accumulation of social capital in the form of inclusive, cooperative networks that are economically productive. Part of the research of Svendsen and Sorenson (2006) was to determine whether the levels of individual perception of productivity and levels of trust would be influenced by measuring social capital as the density of voluntary networking activities.

There are many reasons for getting involved in social networking. Possibly the most important reason is that social media is viral, offering potentially unlimited ability to participate in a connected, secure and information-rich environment that can be tailored to individual needs. Just as important, social media tools are no cost or low cost, allowing a full range of participation in many different situations and organizations.

Any analysis of characteristics of participation must begin with a thorough understanding of the social networking construct. Activities that integrate technology, telecommunications, and social interaction, including the construction of words, pictures, videos, and audio, fit under the social networking umbrella. Social networking is an umbrella term that refers to sharing and discussing information, then using social media, which are primarily Internet- and mobile-based tools. Social media sites are web-based services that allow individuals to use a protected system to construct a public or semi-public profile, maintain a list of people with whom they share a connection, and build and share their list of connections through interactive activities (Brown, 2011). The explosion of new network connections in the workplace suggests an exploration of the various ways managers can affect and improve performance or ways that workers can get better help from their leadership is in order.

Every day, organizations and individuals sign up to take advantage of social networking benefits. But once they are participants, there is an almost inevitable desire to seek information about return on investment of time or effort. The most basic question raised is whether the participant, or the company, is better off because of social networking. Many are quick to accept social networking and the tools it offers. The large and growing numbers of people and organizations involved in social networking suggests that they are, by and large, quick to accept social networking and the tools it offers. Organizations tend to "get in the game" after short periods of deliberation. Once organizations are engaged and leaders decide that an acceptable number of employees are participating, there is inevitably a question posed concerning return on investment. That question may be, "Are my employees better off because we expended money or effort in social networking?" Employees who have limited participation or those who do not participate might ask, "Will my participation make me a more skilled or more productive worker?"

Trusting Relationships

Social networking activities require a trusting atmosphere that allows dependence, satisfaction, and commitment between participants. Trust comes from the ongoing decision to give most people the benefit of the doubt during communication interactions. In the sense of weak ties, trust can even be extended to people one does not know from direct experience. When trust is present, one participant may make a unilateral transfer of control over certain information or resources to another participant, based on the hope or expectation that the other person's actions will satisfy his or her interests. This is an important feature of trust, because one can only be certain about the outcome after the transfer of value has been made.

The independent and interactive effects of trust and dependence on satisfaction and commitment significantly improves organizational effectiveness (Andaleeb, 1996; Schöbel, 2009). Sharing expectations and making inspirational appeals are examples of how leaders can promote trust from the workers. The implied benefit is that workers reciprocate and increase their commitment to the task at hand and, in a larger sense, to the organization, by actively participating in activities.

Improvement Potential

Perceived improvement potential (PIP) involves a conscious determination of whether the activity or commitment involved will bring personal benefits (Brown, 2011). The success of social networking as it relates to performance is based on normative practices as well as on relationships that will build trust, foster interaction, and address participant needs. PIP is based on a self-perception of effectiveness. PIP has a two-fold purpose: to predict the intention to behave in a certain way and to determine whether participants will actually exhibit that behavior.

PIP seeks synergy through integration of network activities. There are three central tasks that are important.

1. Identify the nature and extent of the social relationships characterizing a particular community, its formal institutions, and the interaction between them.
2. Develop institutional strategies based on an understanding of these social relations, particularly the extent of bonding and bridging social capital in a society or a community.

3. Identify ways and means by which positive manifestations of social capital – widespread cooperation, trust institutional efficiency – can offset, and/or be created from, its negative manifestations – sectarianism, isolationism, corruption.

Organizations should find value and cooperation in social networking activities based on satisfaction of the requirements of the central tasks. At this point, aspirations are translated into realities through problem resolution, facilitation of community network building, and increased awareness of how people interact through shared information.

First, social capital allows citizens to resolve collective problems more easily. ... Second, social capital greases the wheels that allow communities to advance smoothly. Where people are trusting and trustworthy, and where they are subject to repeated interactions with fellow citizens, everyday business and social transactions are less costly. ... A third way in which social capital improves our lot is by widening our awareness of the many ways in which our fates are linked. People who have active and trusting connections to others – whether family members, friends, or fellow bowlers – develop or maintain character traits that are good for the rest of society. (Putnam, 1993)

Modeling Opportunities

Organizations can enhance the social networking experience through understanding and employment of effective modeling techniques. Employees can learn social networking behaviors by observing someone else, by examining the outcomes of another employee's participation, or by behavior facilitation. Behavior facilitation occurs when a participant is reminded of a previous learned behavior, or organizational norm, through modeling another participant (Manz & Sims, 1981).

Outcome expectations are also viable for organizational use. This is modeling where participants see others successfully navigating the social networking endeavor. They may even see others being openly recognized for their efforts, and this may create positive expectations. Participants may also gain information from the evidence of a model's behavior that will help form outcome expectancies. Therefore, observing a model is a type of what Manz and Sims (1981) called vicarious learning. Vicarious learning can lead to behavioral change without the learner actually performing the behavior or without directly experiencing the consequences.

Social networking activities can also benefit from social learning theory, which suggests four component processes of observational learning. This theory stresses continuous reciprocal interaction between cognitive, behavioral, and environmental influences (Bandura, 1977):

1. Attention, using modeled events and observer characteristics;
2. Retention, including symbolic coding, cognitive organization, symbolic rehearsal, and motor rehearsal;
3. Motor Reproduction, including physical capabilities, self-observation of reproduction, and accuracy of feedback; and
4. Motivation, including external, vicarious, and self-reinforcement.

There is a very basic approach to modeling that organizations can use that we will refer to as MTT: Model, Train, and Test. MTT requires that organizations model the skill, use training to achieve mastery, and test the new skills in key projects. MTT is based on guided mastery modeling (Bandura, 1986) and is beneficial in addressing task breakdowns and structured reinforcement that coincides with social network interactions. This is how organizations develop intellectual, social, and behavioral competencies.

Begin with modeling the relevant skills for basic competencies. General rules and strategies for dealing with different situations are taught to help people learn how to apply the general rules to specific situations. Once rules and norms are established, provide guidance and create opportunities for skill perfection. This is done in simulated situations where employees can learn without worrying that they might be inadequately skilled or that they will make mistakes. This modeling and guided performance under simulated conditions has been shown to be an excellent way to create competencies (Wood & Bandura, 1989). The skills probably won't be used for a prolonged period unless individuals have a sense that they will be useful skills when put into practice in work situations.

Finally, MTT requires provision of opportunities to find self-directed success. Experiencing a level of success with new-found skills can create high beliefs in both individual abilities and in the new emphasis on social networking activities. It is valuable to allow newly-trained employees to employ their skills in high-level job situations that are likely to produce good results.

Porras (1982) suggested that mastery modeling, MTT for our purposes, can improve not only employee performance and affect morale and productivity in organizations, it can also affect supervisors' skills in a positive way. Supervisors who had the benefit of the modeling program maintained and even improved their supervisory problem-solving skills in ratings provided by their employees. The plant in which the modeling program was applied had a lower absentee rate, lower turnover of employees, and a higher level of productivity in follow-up assessments.

Task Analyzability

Behavioral indicators can prove invaluable to activities that promote new uses of social networking, that seek to increase the current level of participation, or that seek to provide a value proposition that affects participation in a positive way. Elton Mayo's Hawthorne studies from 1924 to 1932 addressed aligning human resources with organizational factors and showed, in part, that workers respond to group norms, social pressures, and observation (Ivancevich, 2008). A group norm is a major factor that the studies have in common with social networking. Task analyzability is a way to allow organizations to take advantage of group norms and other similarities.

When determining how to measure performance, it is useful to start with media richness theory, which is derived from contingency theory (Rice, 1992). The theory proposes that when information processing capabilities match information processing demands, performance will improve. The Rice study examines communication media that differ in media richness or social presence, focusing on conditions of differing task analyzability. The findings show that the affect is on self-reported performance components such as quality of work, effectiveness, productivity, ability to obtain information, decision-making ability, access to others, etc. Rice measured media usage as the reported percentage of information used in one's work that was obtained from online databases.

That brings this examination to task analyzability, which is computed as the mean of four items assessed and standardized by Withey, Daft, and Cooper (1983):

1. "Is there a clearly known way to do the major types of work you normally encounter?"
2. "Is there a clearly defined body of knowledge matter which guides you in doing your work?"
3. "Is there an understandable sequence of steps that can be followed in doing your work?
4. "To do your work, do you actually rely on established procedures and practices?"

Task analyzability requires identification of procedures that define the steps to follow in performing a task. That means building social networking activities into a plan that features well-defined steps that are analyzable and that lend themselves to fit well into the organization's planning activities.

Structured Reinforcement

The next consideration for social networking and improved performance is structured reinforcement. Continuous reinforcement is important because learning experts believe that it is the most important principle of learning (Ivancevich, 2008). A structured reinforcement plan that provides feedback during critical performance times allows real-time information concerning whether the relevant actions are correct or incorrect. Structured reinforcement is characterized by on-demand training, real-time results, ongoing recognition and reinforcement tied to specific actions, and assistance to continue value-plus actions or eliminate value-minus actions.

Structured reinforcement can be accomplished by finding the most appropriate uses of social ties, both weak and strong, and then employing their relevant strengths and weaknesses (Hossain & de Silva, 2009). Weak ties are characterized by absent or infrequent contact, lack of emotional closeness and reciprocal services, and they are considered best suited for innovation (Granovetter, 1973). Strong ties are those that demonstrate strong investment of time and reciprocity. Barry Wellman (1997) argued that sets of actors who maintain strong ties are more likely to trust each other in knowledge sharing, behavior modeling, and in the decision-making process (Wellman & Wortley, 1990). The research suggests that organizations should create as many weak ties as possible to foster open communications and innovation. However, to influence behavior and affect performance, the organization should look to create strong ties.

The bottom line of structured reinforcement is to go beyond simply identifying others who are engaged in the same specialty area and who are taking on the same tasks. Go beyond simply modeling behaviors. The goal is to identify what steps are taken and determine whether the employee's right-now actions are value-plus or value-minus. That type of goal-seeking venture requires more information in the social networking experience and that information can be made available through the use of SNIP.

SNIP is a new source for social networking methodology that focuses on expectation and value, and on return on investment of time in social networking. A foundation is now established to help in defining ways to create a decision matrix, which could lead to improved performance and more effective social networking participation. The matrix uses trust in organizational leadership, PIP, modeling, task analyzability, and structured reinforcement as the basis of a plan of action.

The distinction between social networking and social media should now be clear. Social networking engages online interactions and community-building activities to share experiences, get advice, and make decisions. Social media provides the platforms that allow us to work in those communities.

REFERENCES

Andaleeb, S. S. (1996). An experimental investigation of satisfaction and commitment in marketing channels: The role of trust and dependence. *Journal of Retailing*, *72*(1), 77–93. doi:10.1016/S0022-4359(96)90006-8

Bandura, A. (1977). *Social learning theory*. Englewood Cliffs, NJ: Prentice Hall.

Bandura, A. (1986). *Social foundations of thought and action: a social cognitive theory*. Englewood Cliffs, NJ: Prentice-Hall.

Brown, M. Sr. (2011). *Social networking and individual performance: Examining predictors of participation*. Old Dominion University.

Granovetter, M. S. (1973). The Strength of Weak Ties. *American Journal of Sociology*, *78*(6), 1360–1380. doi:10.1086/225469

Granovetter, M. S. (1982). *The strength of weak ties: A network theory revisited. In Social structure and network analysis* (pp. 105–130). New York: John Wiley and Sons.

Hossain, L., & de Silva, A. (2009). Exploring user acceptance of technology using social networks. *The Journal of High Technology Management Research*, *20*(1), 1–18. doi:10.1016/j.hitech.2009.02.005

Igbaria, M., & Tan, M. (1997). The consequences of information technology acceptance on subsequent individual performance. *Information & Management*, *32*(3), 113–121. doi:10.1016/S0378-7206(97)00006-2

Ivancevich, J. M., Konopaske, R., & Matteson, M. T. (2008). Organizational Behavior and Management. New York: McGraw-Hill/Irwin.

Madden, M., & Jones, S. (2008). *Networked Workers*. Washington, DC: Pew Research Center.

Manz, C. C., & Sims, H. P. Jr. (1981). Vicarious Learning: The Influence of Modeling on Organizational Behavior. *Academy of Management Review*, *6*(1), 105–113.

Porras, J. J., Hargis, K., Patterson, K. J., Maxfield, D. G., Roberts, N., & Bies, R. J. (1982). Modeling-Based Organizational Development: A Longitudinal Assessment. *The Journal of Applied Behavioral Science*, *18*(4), 433–446. doi:10.1177/002188638201800405

Putnam, R. D. (1993). *Bowling Alone*. New York: Simon & Schuster Paperbacks.

Rice, R. E. (1992). Task Analyzability, Use of New Media, and Effectiveness: A Multi-Site Exploration of Media Richness. *Organization Science*, *3*(4), 475–500. doi:10.1287/orsc.3.4.475

Schöbel, M. (2009). Trust in high-reliability organizations. *Social Sciences Information. Information Sur les Sciences Sociales*, *48*(2), 315–333. doi:10.1177/0539018409102416

Svendsen, G., & Sørensen, J. F. L. (2006). The socioeconomic power of social capital. *The International Journal of Sociology and Social Policy*, *26*(9/10), 411–429. doi:10.1108/01443330610690550

Wellman, B. (1997). *An electronic group is virtually a social network. In Culture of the Internet* (pp. 179–205). Hillsdale, NJ: Lawrence Erlbaum.

Wellman, B., & Wortley, S. (1990). Different Strokes from Different Folks: Community Ties and Social Support. *American Journal of Sociology*, *96*(3), 558–588. doi:10.1086/229572

Williams, D. (2006). On and Off the 'Net: Scales for Social Capital in an Online Era. *Journal of Computer-Mediated Communication*, *11*(2), 593–628. doi:10.1111/j.1083-6101.2006.00029.x

Withey, M., Daft, R. L., & Cooper, W. H. (1983). Measures of Perrows Work Unit Technology: An Empirical Assessment and a New Scale. *Academy of Management Journal*, *26*(1), 45–63. doi:10.2307/256134

Wood, R., & Bandura, A. (1989). Social Cognitive Theory of Organizational Management. *Academy of Management Review*, *14*(3), 361–384.

Section 2
Communication Theory

Chapter 5
How Do We Connect?

OVERVIEW

This chapter covers communication methods and a comparison of barriers to communication as they relate to FTF and online interactions. To this point, we have covered a lot of information about social networking, focusing on relevant research and expert testimony. But how do we really connect? To discover the real keys to communication, we have to understand the way communication happens, and then compare and contrast face-to-face and online communications. In looking at various activities in the two approaches, we must understand the existing barriers to communication. Once all of those considerations are covered, we need to understand the steps necessary for full-range communications. For that, we examine models like Michael Argyle's 1972 Communication Cycle, Wilbur Schramm's 1954 Model of Communication that was later revised, and or the Shannon-Weaver Model of Communication. The Shannon-Weaver model demonstrates communication as an interactive process where participants exchange messages and deal with noise in the channel. We address kinesics, proximity, and posture. Kinesics is body language, or communicating through body motions like facial expressions and gestures related to movement of the body in whole or in part. The dictionary defines proxemics as the study of the nature, degree, and effect of the spatial separation that individuals naturally maintain. Western society considers four distances when dealing with relationships between people: intimate, personal, social, and public. We will define those relationships and discuss reading emotions, attitudes, and intentions by analyzing someone's

DOI: 10.4018/978-1-5225-1897-6.ch005

posture. This is an examination of interpersonal communication, which is about defining ourselves through social interaction to create and negotiate identities, relationships, and meanings. Here, we introduce several important references for interpersonal communication discovery and discussion. The first is the second edition of *Engaging Theories in Interpersonal Communication: Multiple Perspectives* by Dawn O. Braithwaite and Paul Schrodt. This work provides great information on a variety of theories that allow a broad examination of the interpersonal communication that is so important to our learning journey. The foreword provides an excellent overview and deep insight by Leslie Baxter, one of the editors for the first edition. The next important reference is a 2004 distinguished scholar article by Leslie A. Baxter titled *Relationships as Dialogues*, which discusses five interrelated conceptions of dialogue. The traditional view of communication is that the process exists to allow people to express their attitudes and beliefs, transmitting them to others so they gain understanding of the sender. In this way, tradition dictates that the sender can influence the attitudes, beliefs, and actions of the receiver to conform to those of the sender. Baxter discusses an alternative, constitutive approach whereby dialogue defines or constructs the social world, which then encompasses and further develops the self and personal relationships. In other words, the dialogue creates an environment in which relationships are formed and people can develop together. Also referenced is an article titled *Communication Theory as a Field* by Roger Craig, portions of which were presented in earlier versions as the Second Annual Lecture in Human Communication, Indiana University, Bloomington, October 16, 1996, and at annual conferences of the International Communication Association, Montreal, May 1997, and the National Communication Association, Chicago, November 1997. This essay examines the field of communication theory, combining the constitutive model with metadiscourse to allow a broader, richer view of human interaction. These important references provide a foundation for the communication discovery we are undertaking here. The chapter concludes with a closer look at barriers to communication and at full-range communication.

Communication Methods

The connection method is basically the same whether face-to-face or online, and we can refer to several communication models as a basis, including Michael Argyle's 1972 Communication Cycle, Wilbur Schramm's 1954 Model of Communication that was later revised, and or the Shannon-Weaver Model of Communication. The Shannon-Weaver model, is shown in Figure 1 and is considered to the "mother of all models." It was created in 1948 when Claude Elwood Shannon wrote an article entitled "A Mathematical Theory of Communication" in *Bell System Technical Journal* with Warren Weaver. This work established communication as an interactive process where the sender and receiver exchange messages and deal with noise in the channel, also referred to as barriers to communication (Sitkin, Sutcliffe et al. 1992, Wagner 1994).

Additional considerations for our communication activities are kinesics, proximity, and posture. Kinesics is body language, or communicating through body motions like facial expressions and gestures related to movement of the body in whole or in part. Gestures, head and hand, or whole body moves can emphasize or reinforce the message. They can relay information about emotions and attitudes or they can conflict with the message that is being sent. Skilled communicators may be able to discover discrepancies and use that information. However, skilled "deceivers" may be able to counteract that activity. The categories of body movement (SkillsYouNeed, 2016) are emblems (serve the same function as a word, such as sticking out a thumb when hitchhiking), illustrators (gestures that accompany words, such as a circular hand movement coupled with the words "over and over"), regulators (head nod to give feedback, for instance), and adaptors (non-verbal behaviors that satisfy a need, such as biting fingernails when nervous).

The Merriam-Webster Dictionary defines proxemics as "the study of the nature, degree, and effect of the spatial separation individuals naturally maintain (as in various social and interpersonal situations) and of how this separation relates to environmental and cultural factors." Each culture has its

Figure 1. Adapted from "A 21st Century Model for Communication in the Global War of Ideas"
(Corman, Tretheway et al. 2007)

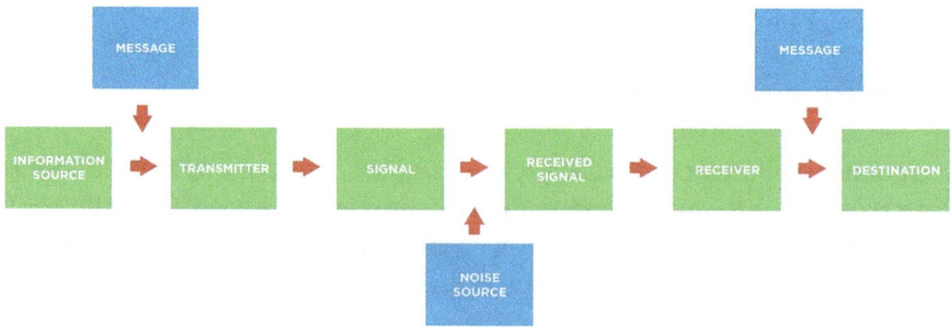

own level of appropriate physical closeness based on types of relationships. People learn theses distances as they grow up in society from parental influence and from societal norms and values. Violating what is considered to be an appropriate distance results in feelings of discomfort and defensiveness in many people. These charged emotions can result in misinterpretations in the communication activity.

Proximity is one of the things we should take into account when communicating and this is why proponents of FTF are adamant about returning to this form over online communication. One study recounted physical structure research that focused on the proximity of organizational members to include the dynamics of relocation (Stryker). The study parameters involved moving organizational members from the traditional closed office to open workstations. In the study, James Stryker says, "the probability of F2F communication between two people is inversely related to the distance separating them." (Stryker) In other words, FTF communication activity has the greatest probability when people are physically close to one another. T.J. Allen reported that the probability of FTF communication between two people is at its highest level when they are located within 10 meters of each other and declines when that distance is greater than 25 meters (Allen, 1977). Our consideration of proximity is relevant because it is the essence of in-person communication, and it can't be used in online communication.

In Western society, four distances are considered when dealing with relationships between people: intimate, personal, social, and public (SkillsYouNeed 2016). Each distance has a close phase and a far phase. Intimate distance is touching to 1.5 feet. This close, we can touch people in intimate ways, we can see the detail of their body, and we can look them in the eyes. At this distance, the two parties tend to ignore or block other people. This is the romance distance. We must remember that we can be threatening when we enter the intimate zone, especially if it comes across as a non-verbal sign that the person can enter the space at will.

The personal zone is 1.5 to 4 feet. Personal zone conversations are very direct. Open and honest good-faith exchanges happen at this distance.

The social zone is 4 to 12 feet and this is where we start to feel a connection with others. We can keep our communication partners at a safe distance, but we can also talk to them without having to shout. A group that is gathered but not necessarily talking directly to each other would be comfortable at this distance.

The public zone is greater than 12 feet. We normally leave this amount of space around us when we're walking around to create a buffer in front of us if possible. This can't always happen, but we take immediate notice when this zone is "broken." We get ready to take appropriate action as people get closer to the social zone. Distance equals safety in our world because it gives us time to see and react to what's coming.

These social distances are estimates and of course are open to discussion. They serve as a basis for understanding human contact. You can get a sense of appropriate social distance by watching people react to each other around you (SkillsYouNeed 2016). Communicators should be aware of these distances and understand the differences, because they allow us to engage others in appropriate and non-threatening interactions. This knowledge can serve to facilitate an understanding of how others view the relationship and how they feel about the distances. Then communicators can make the necessary adjustments.

We can read emotions, attitudes, and intentions by analyzing someone's posture. There are different types of posture, like open and closed, that reflect a person's level of confidence, receptivity, or status in terms of the person with whom they are communicating. Closed position is exhibited by someone who has their legs crossed or arms folded, or who creates a slight angle from the sender. A person sitting directly in front of their communication partner with hands apart on the arms of the chair or on their lap is considered in open position. These positions imply discomfort, disinterest, openness, or interest. Mirroring is a type of posture where people match their stance or attitude to that of their partner. This "copying" of others shows interest and approval between sender and receiver.

Interpersonal Communication

Interpersonal communication is the way we define ourselves through social interaction to create and negotiate identities, relationships, and meanings. The definition of communication has come a long way, moving from just exchanging messages to being defined as a process of sending and receiving messages. Some definitions then morphed to communication dyads until finally we came to view communication as what it really is and should be: a symbolic process humans use to create meaning.

Communication is the way humans build our reality. Human worlds are not made up of objects but of peoples' responses to objects, or their meanings. And these meanings are negotiated in communication. Try not to think of communication as simply a way to share ideas, because it's much more than that. It's the process humans use to define reality itself. (Stewart, 1999)

Human dialogue is the key to interpersonal communication according to Leslie Baxter, a F. Wendell Miller Distinguished Professor of Communication Studies at the University of Iowa. Baxter discusses the importance of the relation between self and other, "a simultaneity of sameness and difference out of which knowing becomes possible. At once, self and other are similar yet different. During interaction, they occupy approximately the same time and space, what is referred to as a chronotope (literally time-space)." (2004) Baxter maintains that individually centered interpersonal communication is about interactions where messages are planned, produced, and processed for shared understanding. One can view interpersonal communication as a search for equilibrium, where the participants engage in error correction to manage the session. There is, and should be, an ongoing movement in message delivery and receipt that emphasizes change and adaptation, thus ensuring message delivery, acceptance, and feedback.

Roger Craig, an associate professor in the Department of Communication at the University of Colorado, discussed metadiscourse in communication. Metadiscourse is about how senders filter ideas based on concerns about how the reader will decode them for understanding. Metadiscourse is important to communication because participants are actively processing information and discovering things like intent and disposition about each other during the exchange. For instance, when one party tells the other, "You don't understand me," there is an urge to seek true understanding by determining why the person said it, what they hope to gain by saying it, and whether they want to discuss it further. This is a form of metadiscourse. This concept refers to the pragmatic use of language to influence the meaning and practical conduct of communication, with more emphasis on the "how" of communication and less on the timeframe or duration.

The frame shift performed by metadiscourse is most often local and momentary, as when a speaker uses the word 'first' to frame an immediate following point as the first in a series of points, or says 'I understand completely' to mark another's statement as understood and accepted. Extended episodes of meta-talk also occur, for example, when a couple sits down to talk over a problem in how they have been talking with each other. (Craig, 2008)

Defining interpersonal communication and understanding dialogue and metadiscourse can give senders improved skills for full-range communications, which begin as soon as we have some information source that helps us form a message. The sender transmits that message as a signal of some kind, verbal or visual, for instance. The receiver accepts the signal and decodes it.

The receiver will then create another message in the form of feedback or a new communication. In a perfect world, this is a collaborative and ongoing process. All communication has some of these elements and actions, but the difference between face-to-face and online interactions is evident in terms of speed and the completeness of the communication.

Let's look more closely at the difference. In face-to-face interaction, we have an idea and we begin to form it to get ready for delivery. Before delivery, we consciously or unconsciously survey the mood and mannerisms of the intended receiver. Are they happy or sad, excited or bored, aggressive or passive? This may not only cause us to delay the message, it may also suggest that we adjust the idea or change it altogether. We try to make eye contact or get some other sign that we have a connection that allows message delivery. When we get to feedback, we are looking for verbal and visual cues that will speak to the effectiveness of our communication interaction. This is drastically different online, because in the digital conversation there is rarely evidence, or time, to suggest a change in the message. Also, there may be inhibitions or barriers based on the human contact, something else that doesn't exist online.

Now let's look at online or social media. The communication happens fast, so we get to deliver it at the time the idea is created, or very shortly thereafter. Many times, in Tweets for example, we distribute an unformed idea or the idea develops as we "talk" in 140 characters at light speed. We've thus moved from idea to formation to communication delivery with no real delay. We hope, of course, that the message is received well and in the way we intended, but are we ever sure that what went out was what we intended? We may even find that the feedback we receive, if we get any, has nothing to do with what we want or need to know about our communication. A great example is when someone posts a short video on Facebook saying that they have come to some conclusions and they are now going to do the right things to make their life better. There can be a rush of family and friends coming online to answer what they perceive to be a distress signal, only to find out that the person has decided to finally take a different route to work to stop getting caught in the same old traffic. I'm sure you have similar examples from your social media activity.

Next, let's look at 11 barriers to communication.

1. Immediacy is the speed at which the communication happens, whether good or bad.

2. Physical barriers refer to equipment used that causes problems. It might be outdated or inadequate, or there might be poor lighting, background noise, or other conditions that hinder communication.
3. Attitudes can interfere with objectivity. Sadness, anger, nervousness, self-righteousness, personal agendas, or dishonesty can hurt communication effectiveness. Some call this "emotional noise."
4. Language can be an issue even if both parties speak the same language. Generational, regional, or country differences can make things difficult, as can colloquialisms, slang, or jargon.
5. Physiological barriers such as pain, poor hearing, poor eyesight, or failing health can hinder communication.
6. Structural Design is about difficulty with understanding how organizations are structured, and problems of supervision or training and poor information systems can make things worse.
7. Cultural Noise can be a problem when stereotypes and assumptions are made based on the receiver's cultural background.
8. Lack of Common Experience can cause problems if the parties involved cannot relate to examples, knowledge, or shared experiences, or when none of these are present.
9. Ambiguity and Abstractions are trouble if there is an abundance of generalizations, points are not clearly made, or the wrong symbols or symbolism are at play. Misinterpretation can occur easily in these cases.
10. Information overload is just what it sounds like; too much information, often with too little time or opportunity to digest the true meaning. Too many details may be bad, so "KISS" (Keep it Simple Sender).
11. Assumptions are dangerous too. The old saying is that when you assume, you make the same thing out of you and me. This problem can also lead a decision to be made before all the facts are presented.

Understanding the barriers is important because they distract receivers and allow senders to get off track. They should be managed well to enhance effective communication. Look at Table 1 for a big picture look at face-to-face vs online interaction.

The chart covers barriers for face-to-face interactions first. First and foremost, face-to-face interactions are two-way communications, which is not always the case with online interactions. In face-to-face, immediacy is high because the sender has verbal and visual cues that he or she must deal with in real time; this is important to successful communication. Physical barriers are high because they have a direct relationship to the success of the communication to allow for receiver comfort. There is a high rating for attitudes

Table 1. Comparing communication styles

Face-to-Face vs Online Interaction		
Activities	**Face-to-Face**	**Online/Social Media**
	Characteristics	
Idea	Depends on receiver interaction	Tends to be fast
Formation	Depends on receiver interaction	Tends to be fast
Communication	Real time flexibility	Short bursts of activity
Feedback	Observe/adjust to meaning: Immediate or delayed	Evaluate meaning: Immediate or missing
Relevance of barriers to message: High/Medium/Low		
Immediacy	Medium to Low	High
Physical Barriers:	High	Low
Attitudes:	High	Medium
Language:	High to Medium	Medium to Low
Physiological Barriers:	High	Low
Structural Design:	Medium	High
Cultural Noise:	High to Medium	Unknown – receiver is often "invisible"
Lack of Common Experience:	Medium	Low
Ambiguity and Abstractions:	High	Low
Information Overload:	Low	High
Assumptions:	High	High

because emotions may have to be dealt with before the message delivery can be completed, or the message may not be delivered at all. Language is high to medium, because the parties can work things out between them and find some understanding during the exchange. Physiological barriers must be addressed to allow attention to the message and feedback. Structural Design is medium because this can be overcome during the communication with the right level of discussion and feedback. Lack of common experience can also be overcome with the cooperation of the parties. Ambiguity and abstractions will be revealed or challenged in the real time of face-to-face interactions, and the parties will have to work through this for success. Information overload tends to be low because the party that is overloaded can simply stop the conversation. Assumptions are rated high because they are problematic and you may or may not have a chance to make up for them.

According to the chart for online interactions, the idea is conceptualized and formed quickly, the communication is accomplished in short bursts of activity, and the sender must evaluate the meaning of any feedback received. That feedback is either immediate or missing based on the platform and the sender's point of view (did I want feedback). In terms of barriers to communication, immediacy is high because the very nature of the digital conversation makes one communicate quickly. Physical barriers are low because you can send and receive from almost anywhere, almost any time, regardless of conditions. Attitudes are listed as medium because the sender can ignore any attitudes he or she doesn't want to deal with. Language is medium to low because there are apps that can take care of any concerns or limitations there; however, the meaning must still come through. Physiological barriers are listed low because they are less of an issue in the online situation because of technological solutions or because the sender can't notice them. I consider cultural noise to be an unknown issue in social media because the receiver is very often invisible or not analyzed. Lack of common experience as well as ambiguity and abstractions are listed as low because, well, they are very often overlooked. In other words, social media is very often "all about me." Finally, I give a high rating to information overload and to assumptions, which are often unchecked in the digital conversation.

The Shannon-Weaver model illustrates that there are several steps involved in communicating that allow participants to exchange messages and receive feedback. Communication exchanges don't always cover all of the sequential steps from the idea of the communication to delivery to understanding and feedback. The communication process shown in Figure 2 takes a different look at the steps that can and should be involved.

Full-range communications require some very definite steps for success. You can look at any number of references, or Google it, or check Wikipedia, but there are some very basic steps you should find in any of these explorations. The eight generally accepted steps are idea development, encoding, message development, communication vehicle selection, message transmission, message receipt, decoding, and feedback. Let's analyze those steps.

1. **Idea Development:** The first step is always having an idea of what you want or need to talk about. Ask yourself: what is the message I want to send? In this step, you come up with a concept that will help you plan your subject and specifics of the communication.
2. **Encoding:** This means translating or converting the intended message into the form you'll use to connect with others. In other words, you might

Figure 2. The communication process

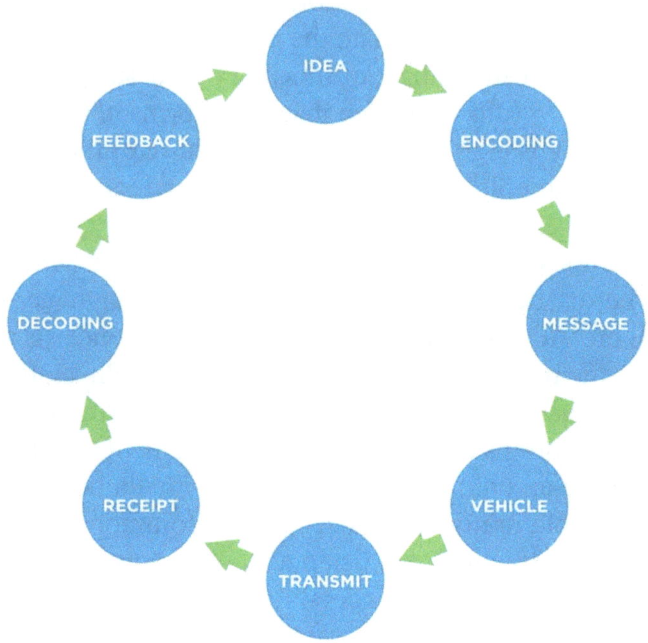

well understand what you mean, but you have to make sure it will be understandable to others.

3. **Message Development:** You might have to change the key words, or the language, or maybe even the positive or negative tone of the message based on your intended audience. This is necessary so that you can transmit the message to a receiver. You'll choose oral, written, or symbolic based on your communication intent. For instance, you might use laughter as a symbolic demonstration of an enjoyable communication.
4. **Communication Vehicle Selection:** The next step is to select the medium or channel. This represents the most effective means of transmitting the message, such as writing, speaking, signaling, or gesturing.
5. **Message Transmission:** This step is simply transmitting the message using the medium you chose in the previous step. This is also the last step in the communication process that is the responsibility of the sender.
6. **Message Receipt:** The receiver's responsibility in the communication cycle begins here. This can be done by seeing the message, hearing the message, or feeling the message.
7. **Decoding:** When the receiver interprets the sender's message it is called decoding. We convert the message into our thoughts as we try to under-

stand and analyze what we've received. For effective communication to occur, both the sender and receiver must assign the same or similar meanings to the message.

8. **Feedback:** This is the final step and it entails the receiver providing a response to the sender's message. This exchange of meaning serves to increase the effectiveness of communication. Feedback signals that the receiver understands the message and it is crucial to effective two-way communication.

We have examined important communication considerations, from theory to process to proximity to interpersonal characteristics. This sets the stage for our next learning stop: comparison of the two communication methods.

REFERENCES

Allen, T. J. (1977). *Managing the flow of technology: technology transfer and the dissemination of technological information within the R&D organization*. Cambridge, MA: MIT Press.

Baxter, L. A. (2004). Relationships as dialogues. *Personal Relationships*, *11*(1), 1–22. doi:10.1111/j.1475-6811.2004.00068.x

Corman, S. R., Tretheway, & Goodall. (2007). *A 21st century model for communication in the global war of ideas*. Consortium for Strategic Communication, Report 701.

Craig, R. T. (2008). *Metadiscourse. In International Encyclopedia of Communication Theory*. Malden, MA: Blackwell.

Sitkin, S., Sutcliffe, K., & Barrios-Choplin, J. (1992). A Dual-Capacity Model of Communication Media Choice in Organizations. *Human Communication Research*, *18*(4), 563–598. doi:10.1111/j.1468-2958.1992.tb00572.x

Skills You Need. (2016). *Body Language, Posture and Proximity*. Retrieved June 12, 2016, from http://www.skillsyouneed.com/ips/body-language.html

Stewart, J. R. (1999).*Bridges not walls: A book about interpersonal communication*. Boston: McGraw Hill College.

Stryker, J. B. (n.d.). Facilitating face-to-face communication in high-tech teams: Face-to-face communication is a necessary component of team collaboration; the physical design of the space, including the visibility of workstations and the availability of community spaces, can affect the level of face-to-face communication. *Research Technology Management, 55*(1), 51.

Wagner, E. D. (1994). In Support of a Functional Definition of Interaction. *American Journal of Distance Education, 8*(2), 6–29. doi:10.1080/08923649409526852

Chapter 6
Comparing FTF and Online Communication Knowledge

OVERVIEW

This chapter brings our knowledge of FTF and online communication together so we can make use of the benefits of each. As stated before, the gap in communication is not about which method works best, it's about bridging the gap between the two. This is a high-touch effort to get all the knowledge in one place. In other words, we have to build the car before we can drive it across the communication bridge. Remember, we have three theories as our foundation: EI, sensemaking, and CAT. Our goal in this chapter is to understand how to apply each theory to both FTF and online communication and then to understand the relative importance of each method. Once we've achieved that, we should be ready and able to move efficiently between the two methods, regardless of the direction of movement. Our hope is that FTF communicators will find ways to keep the human element involved in their online interactions, and online communicators will find a comfort level with leaving the computer to employ verbal and non-verbal to human communication.

Benefits of Face-to-Face Interactions

Face-to-face communication is important because it has to happen whenever humans meet. As hard as we may try to avoid this, it is inevitable in our society. So, even if you're totally online with social media, or you only

DOI: 10.4018/978-1-5225-1897-6.ch006

do virtual reality, there WILL come a time when you must communicate in person. When that happens, there are emotional, physical, and environmental conditions and activities at play. You may choose to ignore them, but they are integral to the interchange and they will affect it, often even if you don't realize it. We start our discussion in this chapter with an examination of the use of theory. Then, we will follow up by discussing the importance of face-to-face interaction and five focus areas that are critical to making it work: comfort, control, confidence, timing, and focus.

Use of Theory

Emotional Intelligence

Using self-awareness and the ability to recognize your own emotions and to have a strong sense of your own tendencies and abilities. Develop people skills to listen actively, acknowledge others, and speak with a purpose.

EI is very useful for face-to-face interactions which require self-awareness and a good feel for recognition of one's own emotions. The sender will benefit from a strong sense of their own tendencies and abilities, and must work to balance those with the receiver's interests. It is absolutely crucial to listen actively, acknowledge others, and speak with a purpose because the receiver is staring right in your face, ready to give you feedback. That feedback will be verbal and non-verbal.

While EI seems tailor made for face-to-face interactions, there is reason for caution. This communication happens in VERY real time and the sender must work to read others, and to anticipate and address their needs. If senders don't verbalize emotions experienced by others and stay in touch with the feelings of both parties, the communication will not be as effective as it can be. But face-to-face interaction gives the sender and receiver multiple chances to build and correct bonds of alignment, appreciation, and support.

Sensemaking

The theory has seven properties:

1. Social context,
2. Personal identity,
3. Retrospect,
4. Salient cues,
5. Ongoing events,

6. Plausibility, and
7. Enactment.

Our focus is on social context and creating meaning from conversations and relationships. Sensemaking is a social activity that requires decoding messages for ourselves and drawing upon a common language we use and on the actions we engage in. It is important to share ideas and influence how others make sense of events.

Face-to-face communications feature great social context, with sender and receiver in a collaborative relationship, creating meaning from conversations and relationships. Even if there is no common language, the parties can work together to find a solution and a way to communicate. Using the full communication process allows sender and receiver to find common ground. Have you ever encountered a person who doesn't speak your language in an airport or other public place? Did you find a way to communicate anyway to get directions or find something to eat? Things like verbal and visual cues allow the parties to communicate.

The sender must assess, early on, the level at which the parties are connected so that he or she can decide whether the activity suits the need to communicate or is the right vehicle for the message. Face-to-face senders must achieve quality early on, or the conversation will end and they won't get to provide "quantity." Because face-to-face senders typically have the whole communication process available, and because they can use the five senses, they can search within the communication for like-mindedness with receivers.

Communication Accommodation Theory (CAT)

Intergroup dynamics are managed by dealing with the various ways in which personal and social identity emerges in social encounters, interactions, and even negotiations. The theory allows examination of how people adjust their speech, vocal patterns, and gestures to accommodate others when they communicate. It examines why people work to emphasize or minimize the differences between themselves and others during non-verbal and verbal communication. The theory also focuses on important interaction considerations, like language, context, identity, and intergroup and interpersonal factors.

Face-to-face communicators need to create social identity and engage in negotiations. The best way to do that is by collaborating well with receivers. They must pay special attention to adjustments within the conversation, adjusting speech, vocal patterns, and gestures to accommodate others. To be truly effective, they must use the full communication cycle as well as the five senses to minimize differences and attend to verbal and non-verbal cues. It is important to work through differences by attending to interaction considerations of language, context, and identity.

Interpersonal control is an issue in face-to-face interactions because the sender wants to influence the receiver's reception and subsequent decoding of the message. The sender gets help with interpretability because there are bits of feedback, visual cues, and emotions being shared by the receiver in real time.

The sender and receiver can work together in discourse management to repair any breaks or difficulties in the communication. Senders must take special care with emotional expression, trying to understand when receivers are expressing, regulating, experiencing, and influencing emotions. But these are clearly observable if not fully understandable, and there's even more help in the interaction in the form of audible sounds, facial movements, and obvious emotional reactions such a laughing, crying, smiling, or scowling. There are unlimited opportunities for the sender to restart, revise, restructure, or reschedule the communication, if only he or she pays attention.

Face-to-Face Communication Importance

Face-to-face communication is important and inevitable, but it may not be the first choice for many, especially Millennials or others who have already switched off their human communication button. There's an obvious impediment to human interaction that is caused by the very presence of cell phones or other technological ways to communicate online.

We see this avoidance of face-to-face communication every day. How often have you been in a room or a public place and seen every person looking at their phone or tablet or laptop instead of a person? In fact, instead of the person or group they're with? The phone is the default communicator now. Many people tend to send a text of one or two words when they could say it to the person sitting right across from them.

What are the reasons people avoid face-to-face communication? There are many, but let's focus on the ones listed in Table 1.

Sometimes, comfort is the reason for avoiding human communication. If I don't want you to answer my question right now, I can text it and delay reading your answer. If I want you to get the information later, I can avoid saying it and then send it when we're apart. If I don't have the answer to your question, I can ask text you that I'm thinking about my answer. Similarly, if I don't have a clue what you're talking about, I can text you to repeat what you said (or texted) so I can Google it and be smarter on my own conditions.

Table 1. Potential FTF issues

Why Avoid FTF?
1. Comfort 2. Control 3. Confidence 4. Timing 5. Focus

So the anxiousness of not having the answer or not understanding what is going on or not wanting to engage can be minimized when the conversation is digital. Finally, it's almost impossible for the other party to see my anger, frustration, shame, or discomfort when the interaction is digital. Conversely, when the communication is face-to-face, verbal, non-verbal, and visual cues can demonstrate discomfort. For instance, does the person start to fidget in their chair, or do they turn their back, or do their eyes start blinking, or do they start stuttering to find an answer? These are ways to discover discomfort with the subject of the communication.

The next consideration is control of the communication. When the interaction is online, I can control when it happens and where I am when it happens. It's very easy to avoid any subject on which I choose not to engage, it's difficult for the other party to challenge me. For instance, how often do you hear "I didn't get your text. I don't know what happened." Isn't that so much easier than the face-to-face "I didn't hear you?" When it's online, the receiver can ask to have the text sent again, giving more time to control the communication. When it's face-to-face, the sender can simply repeat it on the spot. And they're staring you in the face!

Confidence is about the ability of the sender or receiver to achieve a level of certainty that the communication is going the way they want, or that it is about a subject that they want to discuss. You have the opportunity to Google, text someone else, or ask someone in the room with you, without the other party being any the wiser. You can rephrase the expert's words and share them and it seems like you're the scholar. If there's a follow up question, a change in subject, or a request for feedback, you can simply look to your references or act like you never received the online communication. It's all your choice.

The next consideration in our focus is timing, which is easy in online communication. As long as the other party can't see you, you can deny, delay, or avoid the communication. The online communication happens at the time that is best for the sender or receiver, and there's no opportunity to really know if there's a delay with a purpose. When interacting face-to-face, there are verbal, non-verbal, and physical cues that can signal that the other party is uncomfortable or opposed to the timing of the communication.

The fifth consideration on which we focus when using face-to-face communication is the ability, or necessity, to avoid distractions that confuse the communication. In online communication, it is so often about multitasking. The multitasking discussed herein is mental juggling. It happens when you try to perform more than one task simultaneously, perform several tasks in rapid succession, or switch constantly from one task to another. Multitasking is very common in social media communication.

While everyone who uses social media has a favorite such as Facebook, LinkedIn, Twitter, or Instagram, not everyone works exclusively on one platform. Online communicators are bouncing between posting something on Facebook, reading tweets from a friend, viewing the latest images on Instagram, and checking job opportunities on LinkedIn. But every one of those activities takes a part of your consciousness, so you can't focus totally on any single one while all are in play. These communications take the form of constant interruptions that sap you of the ability to recall, repeat, or restructure the communication. Some experts suggest working on social media at a designated time and setting limits on that time to be more focused and productive. But face-to-face communication is right now and to be successful requires the ability to encode and decode before giving feedback. It happens with the other party right there and it happens right now. It also happens with all of the noise in the channel, which are the barriers to communication we discussed earlier. In face-to-face interactions, there is a requirement to manage communication in real time. There's a need to "be in the room."

So we know that face-to-face communication success requires that we pay attention to comfort, control, confidence, timing, and focus. These five areas show us where we must pay attention. Now, we move to online communications.

Digital Communications and Social Media

When we talk about online communication, we are referring to digital communication and social media. The explosion of online interactions is startling and ongoing. Why? Our communications are open to people we can't see, don't know, and have no reason to trust. Yet we communicate things that historically would only be shared with family and friends. We commit dependence, satisfaction, and commitment to strangers, and we tend to respond when they give us feedback. We do this because we make a decision to give most people the benefit of the doubt and we extend that benefit to people we don't know from direct experience. Just like the previous chapter, we start our discussion with an examination of the use of theory. The information presented will demonstrate the importance of social capital and how it enhances online interactions, and will discuss the importance of expectation, value, and return on investment (EVR).

Use of Theory

Emotional Intelligence

Using self-awareness and the ability to recognize your own emotions and to have a strong sense of your own tendencies and abilities. Develop people skills to listen actively, acknowledge others, and speak with a purpose.

EI is very useful for online interactions, since a lot of social media postings can come across as "It's all about me!" Still, the communication activities online allow the sender to focus on and publish their tendencies and abilities. Active listening can happen, but is difficult to judge. But acknowledging others and speaking with a purpose are important parts of this interaction.

The ability or need to read others can be lost in online communication. While the sender might anticipate and address the needs of others, he or she certainly won't have great ability to verbalize emotions experienced by others or stay in touch with their feelings. The sender might want to, but it's very challenging in online communication. This challenge can be helped by the fact that one can work relationship management while online, building bonds of alignment, appreciation, and support.

Sensemaking

The theory has seven properties:

1. Social context,
2. Personal identity,
3. Retrospect,
4. Salient cues,
5. Ongoing events,
6. Plausibility, and
7. Enactment.

Our focus is on social context and creating meaning from conversations and relationships. Sensemaking is a social activity that requires decoding messages for ourselves and drawing upon a common language we use and upon the actions we engage in. It is important to share ideas and influence how others make sense of events.

Online communications certainly create meaning from conversations and relationships in true social context fashion. The limitations at work here have to do with decoding messages. The parties don't always have a common language and, without other parts of the communication process, that understanding and common ground might not be reached. More startling is the fact that the communication almost always continues without it. That actuality can taint future communication activities and make it difficult to share information.

As with all communications, it's important to understand the level at which you're connected and decide whether it suits your needs. This can be a challenge in online communications when your audience may or may not answer or acknowledge you. Social media bloggers must decide on quantity versus quality in relationships. What balance is reasonable and manageable? Many people say they "know" the people they deal with on social media, but maybe not. Will you recognize them when their picture is an eggshell or a dog or a building? Can you stop by their home for dinner? Do you know where they live? Finally, online interactions make it difficult to search for like-mindedness, which often gets lost in the shuffle in the large number of impersonal Internet communications.

Communication Accommodation Theory (CAT)

Intergroup dynamics are managed by dealing with various ways personal and social identity emerges in social encounters, interactions and even negotiations. The theory allows examination of how people adjust their speech, vocal patterns, and gestures to accommodate others when they communicate. It examines why people work to emphasize or minimize the differences between themselves and others during non-verbal and verbal communication. The theory also focuses on important interaction considerations, like language, context, identity, and intergroup and interpersonal factors.

Communicators in online activities work hard to enable social identity and even to engage in negotiations. There is an emphasis on adjusting to the conversation and even attempts to adjust speech or what is called voice in social media, but this method cannot truly adjust vocal patterns and gestures to accommodate others. Differences may be minimized, but there is a lack of non-verbal communication assistance. Language and context can be managed, but identity brings with it imposters or others who conduct unethical behavior.

There are online communication issues where CAT is concerned. Interpersonal control is an issue because the sender does not have a lot of influence over the receiver; the power is all in the hands of the receiver. Interpretability could be helped with feedback, but feedback doesn't always come to the online communicator. Interpretability could be helped with visual cues, emotions, or other characteristics of the receiver, but they are not always readily available in the online space. One exception is online video chat or collaboration.

Discourse management is necessary to repair any difficulties or breaks in the communication, but without real contact it is a challenge to direct the conversation or communication because there is not a sufficient level of feedback to help in this matter. Emotional expression is about people expressing, regulating, experiencing, and influencing emotions. Emotional expressions are not observable (except in online chat) because they take the form of verbal and non-verbal behaviors. That means we are talking about audible sounds and facial movements in addition to obvious emotional reactions such a laughing, crying, smiling, or scowling. All of these activities are difficult or impossible in online interactions.

Online Connections and Social Capital

The online conversation is important, because it allows us to connect with the environment and impact social welfare. It puts us in touch with people and organizations that we might never otherwise discover. These activities and the technology that allows them are cutting-edge approaches to communicate in real time. To be effective, and these interactions can be, they must focus on the value proposition that is at stake and maximize the effort to promote the best in communication.

Online communications require attention to social capital theory, which suggests that the efficiency of society can be improved by facilitating coordinated actions. While it's easy to participate in the digital conversation without a second thought, it's important to consider actions that deliver meaningful messages, foster two-way understanding, and create feedback that enhances the next communication steps.

Social capital can illustrate how active and trusting connections with others widen our awareness and reveal commonalities in society that often prove beneficial. Social media allows us to arrive at these outcomes. Social capital is about establishing relationships purposefully and employing them to generate intangible and tangible social, psychological, emotional, and economic benefits in short or long terms. Social capital can be examined in terms of five dimensions: networks or lateral associations between individuals and/

or groups, reciprocity and expectation, trust and risk based on assumptions, social norms, and personal and collective efficacy.

While the social capital conversation here is individually focused, there are also organizational benefits available. There are many reasons why corporations bring people together online and use social networking to integrate virtual communities into existing business models. Companies are interested in social networking because of their potential to deliver measurable benefits like innovation, more effective marketing, more rapid access to information, lower costs and higher revenues. Corporations may find compelling benefits through engagement, which is the conscious attempt to use interaction to gain attention and reputation. The company begins with an understanding of what knowledge can be shared that is worthy of giving and receiving input. Social media is then used to create compelling content, foster adoption or participation, filter incoming and outgoing information, and maintain levels of activity. Companies also will determine appropriate online spaces and channels, and then set goals and milestones to develop participation opportunities that will last.

Effective online interaction allows companies to benefit by engaging appropriately to gain adoption and trust within the community of interest. Adoption should be a by-product of good bookmarking and tagging practices, and of adding people to social media groups. Attention and reputation should reasonably offer continuing collaboration activities.

The attention to social capital is important because the speed at which online communication happens can make it very difficult to conduct full-range interactions. What that means is that the idea may not be properly or completely formed (encoding) or the receiver may not have enough background or other information to properly decipher the intended meaning (decoding). An even bigger problem is that without the use of the five senses to get feedback on the message, the parties involved can come away with what they think is understanding, but it could actually be two distinctly different messages; one sent and another one received. When that happens, there is a danger that the next activity will be faulty, because the basis is now an incorrect assumption.

One way to address these communication issues is to take advantage of all that social media has to offer by leveraging the expectation, value, and return on investment (EVR). EVR is about identifying, achieving, and nurturing social capital, which is the value or reward for effectively applying the social media strategy.

EVR addresses goals, establishes roles, and seeks feedback and verification that the online communication is working. EVR has several objectives. First, identify and improve social media participation by delivering on expectations of shared value and a worthy return on each company's, or participant's, investment of time and attention. Second, develop or improve organizational or personal brand using a two-way communication arrangement where valuable information is exchanged freely, and where giving and getting recognition matters to the participants. Third, develop a social media plan that addresses the goals required by the company or personal brand.

In this way, communication effectiveness relies on understanding the value proposition of the interaction. So online interactions can be enhanced by focusing on value propositions of the activity, and tracking or measuring progress can only lead to improved participation and results.

Online communications are very effective, fast, and to the point. No one would argue that there is more communication than there needs to be at times, but the benefits are there. The premise of this book is to highlight what's missing. Online communications often leave out important elements of the communication process. Whether that is for expediency or comfort, it causes problems not only in the completeness, accuracy, and efficiency of the interaction, but also in the inevitable need to do some measure of face-to-face communications.

In this chapter and the previous one, looking closely at the primary theories for this book helps illustrate important considerations. The next consideration is a "deep dive" into interpersonal communication to fuel our educational experience.

Chapter 7
Interpersonal and Group Communication

OVERVIEW

In this chapter, we begin with a different theoretical examination and then finish with an overview of why we communicate. We will focus on interpersonal communication theories to continue the learning venture, and then cover group communication. Communication theory was covered in Chapters 3 and 6, but now we turn to additional discovery relevant to interpersonal communication. Six theories will be relevant for our discussion. Uncertainty Reduction Theory tells us that people's uncertainty with those they don't know can motivate them to communicate to reduce uncertainty. Supportive Communication Theory is defined as "verbal and non-verbal behavior produced with the intention of providing assistance to others perceived as needing that aid. Interpersonal Deception Theory states that senders try to manipulate messages to be untruthful causing apprehension on the part of the sender due to the concern that their false communication will be detected. The motivated information theory is about managing information management activities in relation to interpersonal communication, combining a three-phase process – interpretation, evaluation, and decision stages – that individuals use to seek or avoid information about an issue. Communication privacy management theory addresses important ways that people process information. It supposes that people have private information, private boundaries, control and ownership about the information, and we control it with rule based management and privacy management dialectics. Affection exchange theory focuses on affectionate communication as an adaptive behavior that contributes to hu-

DOI: 10.4018/978-1-5225-1897-6.ch007

mans' long-term interaction and success. As we focus our social interaction, we must understand that direct interpersonal communication is a face-to-face relationship between the parties involved in communication, and this is an interdependent arrangement. Feedback is a crucial component because of its immediacy and primacy. We examine the resolution of conflict to save face, correct the record, determine winners or losers, or to set terms for future interactions. This is important because conflict is often studied at the interpersonal level and it is often a part of communication. We then cover group communication by providing a theoretical backdrop for behavior in groups, first using research to describe individualistic and collectivistic mindsets. That is followed by a description and analysis of three theories. Aubrey Fisher's Decision Emergence Theory demonstrates how groups progress through stages to address diversity leading to shared experiences, decision methods, work ethic, and values. Forming, Storming, Norming, and Performing are parts of a process that is about stages that move groups through task achievement and member satisfaction, while dealing with in-group conflict as it arises. Group interaction in terms of standards, power, meaning, and values is addressed using Structuration Theory. This examination of relevant theory is somewhat different from what we discussed earlier in the book.

Uncertainty Reduction Theory

Uncertainty Reduction Theory, Supportive Communication Theory, Interpersonal Deception Theory, Affection Exchange Theory, Communication Privacy Management Theory, and the Theory of Motivated Information Management will assist the current discussion of FTF and online communication activities. Uncertainty Reduction Theory holds that people feel uncertainty about those they don't know and are motivated to communicate to reduce uncertainty. The theory was invented by Professors Charles Berger and Richard Calebrese in 1975. They discovered that initial interactions between individuals follow predictable patterns of information gathering (1975). Relationship development engages uncertainty reduction because individual behavior can be predicted using information gathered through observation. There are several assumptions in uncertainty reduction theory, the primary one being that uncertainty creates cognitive discomfort that people automatically try to reduce. The process of information seeking goes through predictable developmental stages, indicating changes in the quantity and type of information shared between individuals. Berger and Calabrese outlined seven concepts related to these assumptions (1975):

1. **Verbal Output:** High levels of verbal output correlate positively with a greater reduction of uncertainty, higher levels of communication intimacy, similarity between individuals and liking.
2. **Non-Verbal Warmth:** Refers to positive signs in a person's gestures and body language that indicate a willingness to communicate or form a relationship.
3. **Information Seeking:** Occurs when individuals wish to know more about each other. Information can be obtained passively through observation or interactively through conversation.
4. **Self-Disclosure:** Individuals willingly divulge information about themselves to reduce uncertainty in the other person, thus encouraging them to communicate openly.
5. **Reciprocity:** Individuals interested in reducing uncertainty or starting a relationship will reciprocate uncertainty-reducing behavior, such as asking questions. The higher the uncertainty between individuals, the more reciprocity a person can expect.
6. **Similarity:** Individuals who are alike or share interests will feel less uncertain about each other and achieve communication intimacy more quickly. Dissimilar individuals experience higher levels of uncertainty.
7. **Liking:** Feelings of approval and preference between individuals likewise speed up the uncertainty-reduction process. Feelings of dislike discourage relationship formation.

Supportive Communication Theory

Supportive Communication Theory is defined as "verbal and non-verbal behavior produced with the intention of providing assistance to others perceived as needing that aid." (Burleson & MacGeorge 2002) Social networks are created through supportive messages, but there's more to it. There are also social experiences and psychological dispositions that occur, and the messages may deliver additional benefits in decreased emotional distress, enhanced coping, protected health and enhanced relationships. These benefits are indicative of the value available through good communication. The sender's supportive message is intended to help the receiver in some way, and the message contains verbal content, non-content features, paraverbal features, and non-verbal features (Burleson 2009). The explicit utterings of the sender and the logical and pragmatic presuppositions and implications of these statements make up the verbal content of the message. Messages should be examined for their global focus, which is the extent to with the message focuses on problems versus emotions and the extent to which the message engages or avoids receiver outcomes (Barbee & Cunningham 1995). Senders should also consider politeness or stylistic approaches that attempt to minimize the effects of evaluations and impositions, the type of support that is intended by the message (emotional, informational, advice, esteem, or social network), and the quality of support. In terms of quality, ensure that the message is understandable and relevant to the issue at hand, that it has a good chance to solve or improve the issue, and that it is achievable.

Non-content features are the second feature of supportive messaging. These features include the length of the message, the number of appeals, statements or strategies in the message, and the timing or sequential placement of message content in the interaction (Jacobson 1986, Neff & Karney 2005; Feng 2009). Paraverbal features of the message include the rate, pitch, and volume of the delivery, hesitations, pauses, and silences during delivery, and the overall smoothness or fluency of delivery (Dolin & Booth-Butterfield 1993; Bullis & Horn 1995). The fourth feature, non-verbal aspects of the message, relates to immediacy, involvement, pleasantness, eye contact, and smiling can produce desirable recipient responses like enhanced task performance or improved affect.

Interpersonal Deception Theory

In the late 1990s, Judee Burgoon and David Buller introduced Interpersonal Deception Theory, which states that senders try to manipulate messages to be untruthful (Buller and Burgoon 1996). This causes apprehension on the part of the sender due to the concern that their false communication will be detected. At the same time, receivers try to determine the validity of the information, creating suspicion about whether the sender is being deceitful. Deceptive messages have three parts:

1. The central deceptive message which is usually verbal,
2. The ancillary message containing verbal and non-verbal communication, often revealing how truthful the message might be, and
3. Inadvertent behaviors, which are primarily non-verbal and help demonstrate the deceit of the sender through a concept known as leakage.

Leakage occurs when non-verbal signals betray the true content of a contradictory verbal message. Facial expressions may signal deception, but they are sometimes hard to read. Non-verbal signals for deception include the following:

- Pronounced hesitations
- Pupil dilation
- Increased rate of speech or tempo
- Increased, rapid, or strained blinking
- Voice changes pitch
- Sweating

Theory of Motivated Information Management

This theory was introduced in 2004 by Walid Affifi and Judith Weiner, and then revised by Afifi and Christopher Morse in 2009. "Employees turn to managers for performance feedback, students rely on instructors for information about classroom material, individuals seek health information from physicians, and partners depend on each other for relational information." (Afifi & Weiner 2004) The theory is about understanding information management activities in relation to interpersonal communication, combining a three-phase process – interpretation, evaluation, and decision stages – that individuals use to seek or avoid information about an issue.

The interpretation phase is characterized by awareness of an uncertainty discrepancy about an important issue and the anxiety that ensues, the evaluation phase reflects expectations about the outcomes of an information search and perceived abilities associated with that decision, and the decision phase involves the selection of information-management strategies. (Afifi & Weiner 2004)

There is a similar, two-stage process – a breakdown of information seeker and information provider roles – which senders use to decide what, if any, information to provide. The theory demonstrates an iterative process that assesses the impact of how much the information provider will share and how they share it. Information providers and information seekers go through the same evaluation and decision phases, but the information provider makes an outcome assessment to decide the pros and cons of giving the seeker the desired information. The relationship between information provider and information seeker, or sender and receiver, is consistent with the feedback loop traditionally shown in most communication models. Both participants in the interaction predict the outcome of revealing the information, determine the importance of that outcome, and assess the probability that the providing the information will result in the anticipated outcome. All of the communication components are weighed in combination, and communication assessments are made along coping, communication, and target lines.

Communication Privacy Management Theory

Communication Privacy Management Theory is about the decisions we make every day and about how much information we are willing disclose to people and how much we want to keep private. There are risks and rewards in terms of the amount of information we disclose or keep private. We can think of this theory as a way to navigate the road to privacy through boundaries that surround information we have but others don't. This theory has several suppositions: people have private information, private boundaries, control and ownership about the information, and we control it with rule based management and privacy management dialectics. The theory is based on 5 core principles (Petronio 2015):

1. People believe they own and have a right to control their private information.
2. People control their private information through the use of personal privacy rules.
3. When others are told or given access to a person's private information, they become co-owners of that information.
4. Co-owners of private information need to negotiate mutually agreeable privacy rules about telling others.
5. When co-owners of private information don't effectively negotiate and follow mutually held privacy rules, boundary turbulence is likely to result.

Affection Exchange Theory

Affection Exchange Theory focuses on affectionate communication as an adaptive behavior that contributes to humans' long-term interaction and success (Floyd 2015). The theory demonstrates that affection exchange is a major factor in survival because it promotes bonding between people. Like other forms of communication, affection exchange communication involves verbal and non-verbal messages that convey "feelings of fondness, support, and love."

The theories presented here provide insight and resources for the continuing communication discussion.

Focused Social Interaction

Interpersonal communication is most effective when social interaction is focused to build a communicative relationship with the parties involved. Direct interpersonal communication is a face-to-face relationship between the parties involved in communication, and this is an interdependent arrangement. The communication activity has a strong feedback component, even if it is not taken advantage of, because of its immediacy and primacy (in the right now and right here). Relationships that last over long periods of time happen because of enhanced communication. Effective interpersonal interaction is dedicated to the use of the words used as well as the various elements of non-verbal communication.

Figure 1 highlights Interpersonal Communication, which is defining ourselves through social interaction to create and negotiate identities, relationships, and meanings.

Interpersonal communicators seek outcomes such as help, discovery, influence, and sharing. We can also distinguish between interpersonal and impersonal communication. "Impersonal communication is that which involves functional short-term exchanges such as might occur between a shopper and a salesman; the label of interpersonal is reserved for communication that functions in deeper and more meaningful relationships." (Knapp 1983)

Figure 1.
© Can Stock Photo Inc./michaeldb.

Communication scholar Mark Knapp believes that there are several stages of interpersonal communication that occur over an extended life cycle, and he outlines a framework for this important process (1983). The stages are:

- **Initial Encounter:** This is a data-rich first impression. Likes or dislikes can be instantaneous, though many people have learned that first impressions may be misleading.
- **Experimenting:** This second step is characterized by information exchange on a variety of topics. This step is at first generic and safe, but it can gradually become more probing, self-revealing, and personal.
- **Intensifying:** Positive mutual conclusions are reached based on experimentation that shows the relationship is worth pursuing.
- **Integration:** In this stage of intense friendships, close business partnerships, romantic commitments, etc., mutual decisions can be made to agree that the relationship is fulfilling.
- **Bonding:** Here the relationship is sealed and publicized, characterized by contracts, written agreements, and announcements.

Knapp also outlined a similar reverse pattern for the unraveling of interpersonal relationships:

- **Differentiating:** Similar to initiation, but this focuses on the differences that people notice about each other.
- **Circumscribing:** Interactions are minimized and the conversation turns primarily to functional topics.
- **Stagnating:** The relationship struggles and is unfulfilling, but is continued for external reasons such as religious or family obligations, contractual obligations, or social expectations.
- **Avoidance:** Characterized by overt unpleasantness, participants avoid each other and express mutual annoyance when they encounter each other.
- **Termination:** This stage is signaled by the termination of legal, religious, or other formal contracts and the participants publicly announce the end of the relationship.

Knowledge of the communication stages is important, but there are interpersonal communication barriers that begin with encoding problems for the sender and decoding biases for the receiver. Senders can have difficulty if they choose the wrong medium. Senders must understand the methods and correctly choose verbal, written, or non-verbal means for delivery. Verbal communication is a way to provide immediate feedback, and it is best for demonstrating emotions and building interpersonal relationships. Policy-setting conversations and storytelling are verbal examples. Written communications facilitate appropriate use of grammar, word choice, structure, and punctuation for effective messaging. Visual cues allow the sending of non-verbal communication. Communication can also suffer from bad timing, for instance, delivering a complicated message when the receiver is pressed for time or trying to use humor in an emotionally-charged interaction. Failure to read feedback can always cause problems for both sender and receiver.

Dealing with Conflict

Next, conflict is often studied at the interpersonal level because it is such a vital part of communication. Interpersonal communication often requires resolution of conflict to save face, correct the record, determine winners or losers, or to set terms for future interactions. To be an effective communicator, one must be good at conflict resolution, so let's use Table 1 to examine some important skills. Senders need to have an ability for consensus building, which requires negotiating the creation of a mutually acceptable agreement leading to an agreed upon conclusion. The consensus must provide suitable accommodation to some degree for all participants.

Conflict management is important in difficult interactions, and it requires anticipation, research, and identification of potential solutions that lead to problem solution. The primary focus of conflict management, which is more an organizational than a personal concern, is to change conditions without addressing the internal or external cause of the conflict. Conflict resolution is finding a permanent solution for a problem or dispute, and it is achieved through dialogue without physical or verbal violence.

For our personal discussion of communication, we use adjusted conflict styles listed in the table below to make the organizational approach a personal one. Integrating and compromising styles are focused on solving the

Table 1. Understanding conflict styles

Adjusted Conflict Styles
1. Integrating style involves a high level of concern for oneself and for others.
2. Compromising style involves a moderate level of concern for both parties.
3. Dominating style involves a high concern for oneself and a low concern for others.
4. Obliging style stems from a low concern for oneself and a high concern for others.
5. Avoiding style involves a low concern for both.

conflict. Dominating style is focused less on problem solving and more on control. Obliging and avoiding styles are about trying to avoid the confrontation altogether.

Conflict happens, but it can be positive or negative. It is positive when communication is increased, stored feelings are released, problems are solved, performance is improved, or the relationship grows. Negative conflict is at play when parties don't address real issues or problems, when people become uncooperative, or the relationship between the parties is damaged or disrupted. Conflict is sometimes categorized either as expressive conflict, rooted in a desire to release tension, or as instrumental conflict that stems from a difference in goals or practices (Knapp 1983).

More specifically, good interpersonal communication is also important when the interaction is toxic. What's a toxic conversation? The message may be misunderstood at the beginning, but the receiver may view the activity as verbal warfare. When this happens, the whole interaction can be complicated by subconscious, unspoken intentions and perceptions. The use of verbal and visual cues is absolutely critical when the parties are confused or shocked or may be feeling doubts about continuing the interaction. In *Failure to Communicate*, Holly Weeks dissects hard talks (toxic conversation) to identify the challenge and set the stage for improving conversations (2008).

The three basic traits of hard talks are combat mentality, emotional loads, and hard to read. Combat mentality happens when one or both parties see the interaction as a conflict that needs a winner and a loser. Hard talks carry much heavier emotional loads than normal ones, and these emotions are often anger, embarrassment, anxiety, or fear. The third trait is when the other person's intentions are hard to read, or when you can't tell how they are taking what is said or how they might feel about it or react to it. Weeks says that people bring three conventional ideas to tough conversations, ideas that are not hard to think about:

1. Tough conversations are simple,
2. We can win them, and
3. They aren't our fault (2008).

These ideas are misguided and they don't help us, primarily due to oversimplification of the problem and the win-or-lose mentality. First of all, tough conversations are hard because there are emotions or perceptions at play, and they must be addressed or, at least, acknowledged. Second, the idea of winning or losing is counterproductive. What we want to accomplish is a compromise or collaboration where both parties get something they value. We want the "win-win." Third, "they aren't our fault" is simply avoiding the issue. In great communication, in any communication, we want buy-in from everyone. If both parties can take ownership of the issue, then they can work together to resolve it. Again, compromise or collaboration is what we're looking for.

Interpersonal communication is a shared responsibility issue, and that is definitely true with the hard talks or toxic conversations. When dealing with the combat mentality, it is important to bypass the urge to force winning and losing, and focus on strategy and skill. This requires making sure that all communication intentions and plans include the knowledge of where we want to go with the message and an assumption that there will be difficulties at some point in the interaction. Strategic thinking is important when dealing with communication combativeness. Strategic thinking is the ability to use your communication skills conceptually, imaginatively, systematically, and opportunistically to achieve success. When dealing with combative communications, Weeks suggests focusing on a preferred outcome to arrive at a preferred relationship, dealing with where we want to get, where we think we can get, and where we're trying to get in terms of our message delivery (2008). Communicators also want to identify interference in the channel, or what we call barriers to communication. The combative issues can also be addressed with three-way respect: respect for our counterpart, self-respect, and respect for the problem between us (Weeks 2008). This is truly an interactive solution: high touch!

When the toxic conversation is hindered by emotional loads, we must use our skill to navigate the difficulty. The first thing to do is to manage your own emotions so they don't add an unreasonable burden to the interaction. If you're too emotional to continue, change the communication approach or reschedule the communication. Changing the approach could be moving to the next message or restating the message to soften its impact. Next, recognize the power of the emotion and then find a middle ground, a place where the conversation can continue. Communicators can also try to refocus the conversation on the intent and the end state, hopefully diffusing the emotional load.

The previous discussion dealing with emotional loads lends itself to an application of EI, emotional intelligence. EI can be very valuable in achieving success in interpersonal communication activities. It is helpful to draw on two of the five EI components – empathy and social expertness – for our interpersonal communication discussion. Empathy is the foundation for building relationships. "Empathetic people genuinely care about other people and place a high value on understanding what is going on in the hearts and minds of others." (Wall 2008) That understanding is the cornerstone of interpersonal communication. When you send a message, you want the receiver to decode it, care about it, and give you feedback. Do whatever you can to properly encode the message and make it appealing. Senders should work to engage verbal and non-verbal cues to enhance empathetic actions. First listen, then put into your own words what people are saying to enhance understanding.

Social expertness in EI means building relationships to get things done. People and their interactions can get things done if healthy relationships exist. Strong relationships are characterized by collaboration, feedback, and mutual trust. In *Working Relationships: Using Emotional Intelligence to Enhance Your Effectiveness With Others*, Bob Wall offers a sequence of steps that leaders in organizations can use to gain useful feedback (Wall 2008). This organizational focus is useful for individuals interested in effective two-way conversations. Table 2 provides an abbreviated version.

The third trait of hard talks is called hard to read, and the way to overcome this roadblock is more art than skill. This takes relationship building to get it right. Don't avoid dealing with what is going wrong in the conversation and help the receiver do the same. If you haven't stated and agreed on what each party intends and what they think to be the focus of the communication, do it now. If you've already stated it, restate to be sure both sides are on the same page. This is important to allow each party to avoid confusion, tension and suspicion that could lead to collapse of respect.

Interpersonal communication and its characteristics are at play in group communication, with somewhat different considerations and expected outcomes.

Group Communication

We begin our group communication discussion with information about group behavior. There is an abundance of evidence in management research that teaches about individualistic versus collectivistic mindsets (Wagner 1995;

Table 2. Ways to seek feedback

Asking for Feedback
✓ Schedule the conversation at a convenient time in a comfortable place.
✓ Establish your purpose.
✓ Put the other person at ease.
✓ Ask open-ended questions.
✓ Relax and listen.
✓ Ask for examples to clarify broad generalizations.
✓ Do not defend yourself.
✓ Close the conversation with gratitude.
✓ Take action.

Chatman, Polzer et al. 1998; Kanter 2000; Locke, Tirnauer et al. 2001). Management scholars explain that those with an individualistic mindset put their own needs above everyone else. Those with a collectivistic mindset, on the other hand, put the needs of the group before their own. In individualism, which is traditionally matched to the American workplace, the person sees their life as their own, acts on their own judgment without regard to others, and pursues whatever values they choose. An individualistic person focuses on me, myself and I, holding his or her own opinion as more important than the group's and emphasizing task over relationship. This person is interested in relationships, but not at the expense of a task they value. This can, of course, be counterproductive to group communication or activities if there are too many members with this point of view. Consensus becomes tedious of not impossible.

The collectivistic mindset is considered conducive to good group operations, allowing motivation, reducing social loafing, increasing cooperation, and promoting feelings of harmony. The down side is that the group's desire to cooperate may overcome the need to be creative enough to produce innovation. The notion of collectivism is that people should sacrifice to achieve the best outcome for the group. A collectivistic person focuses on us versus them and believes the group opinion is paramount in importance. In this view, it can be difficult to get direct answers from people. There is a fear of being embarrassed or "losing face," and the group relationship takes precedence over the task.

These positions are based on Geert Hofstede's dimensions of national cultures which distinguishes countries from each other by grouping them statistically into four clusters to deal with anthropological problem areas: ways of coping with inequality, ways of coping with uncertainty, the relationship of the primary group to an individual member, and the emotional implications of gender at birth (1991). These areas then became Hofstede's dimensions, which are power distance, uncertainty avoidance, individualism versus collectivism, and masculinity versus femininity. It is possible to view group communication in terms of the individualism-collectivism orientation based on Hofstede's principles. The cultural position of group members in terms of these dimensions gives an indication of how to get group consensus. If we can determine which members are operating from each viewpoint, we can affect consensus. Individualistic-oriented people tend to value the concepts of autonomy, self-interest, and performance. Collectivistic-oriented people exhibit a higher level of emotional dependence on the team, and are more conforming, orderly, traditional, and team-oriented (Matveev & Nelson 2004).

This discussion leads us to investigate how group dynamics work. A number of theories attempt to analyze and explain the dynamics of group communication and the elements that determine its effectiveness. We will focus on Decision Emergence Theory, the theory of Forming, Storming, Norming and Performing, and Structuration Theory.

The first part of this analysis is on Aubrey Fisher's Decision Emergence Theory, which states that new groups progress through four stages that lead the group from a diverse collection of individuals to a productive unit with shared experiences, decision methods, work ethic, and values (Fisher & Ellis 1980). This theory works best when conflict is low. The four stages are orientation, conflict, emergence, and reinforcement. Orientation features team members building relationships amid tensions. Senders must take special care in this stage, because the communication is difficult due to the fact that bonds have not been formed and cannot exist at this stage. The conflict stage is where the group discusses new ideas. There is usually a significant amount of tension during these early deliberations as members of the group work to win or accept roles. Care must be taken to ensure that the group stays together in this stage.

The outcome of the conflict stage comes into play in the emergence stage. This stage requires that team members be willing to adjust their positions to avoid the perception of dominance. Here is where the needs of the group must outweigh the needs of the individual. Team members need to show commitment to plans and objectives in the fourth stage, emergence. Group success requires commitment even if some members do not personally agree. But they must put personal issues aside for the good of the group.

Fisher's work is related to another theory, Forming, Storming, Norming and Performing (Tuckman & Jensen 1977; Smith 2005). This view, illustrated in Figure 2, is about the stages through which groups progress. Bruce Tuckman developed this knowledge with four stages, then added a fifth in the 1970s. The knowledge is based on people wanting information about other members of the group, and the stages help the group achieve tasks and satisfy members. It is also about dealing with conflict as it arises. Forming is the stage where groups orient and test themselves, looking to set interpersonal and task behavior boundaries. Then, the group establishes relationships of dependency with leaders, other group members, or pre-existing standards. Forming is the very beginning of group activities, so typically there is very little consensus on what should be accomplished. The group is very dependent on the leader at this point, especially since that is often the only role that has been decided or assigned. There are lots of questions at this point.

Figure 2. Stages of group development

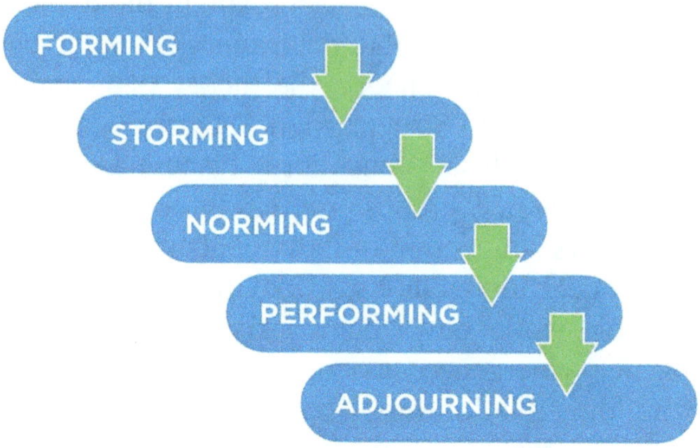

Next, groups go through storming where they deal with interpersonal issues which can cause conflict and polarization. Groups need to watch emotions in this stage as they work against, or through, group influence and task requirements. Storming is dynamic, because team members are jockeying for positions and it's hard to get a consensus. There is more clarity as people start to assume their roles, but this is when factions and cliques form. The group must overcome power struggles and avoid getting sidetracked with emotional and relationship issues. Compromise is needed and questions come sporadically and without focus.

Norming is the third stage, and this is where resistance is conquered, allowing bonding and cohesiveness to occur. This is also where groups establish standards and roles, and where members express their personal opinions, sometimes in a very strong fashion. But the group is coming together now. Consensus and agreement are formed in the norming stage with roles defined and accepted and good response to the leader by group members. The group is productive, making big decisions and possibly delegating small decisions to members or small teams. The team functions more as a family or social club. The interaction goes beyond questioning to sharing and collaboration.

The fourth stage is performing, and now the group uses the interpersonal interactions to move the group's energy toward task completion. Roles are now functional and flexible and structural issues have been resolved. Performing is characterized by a group that knows what it wants to do and why it wants to do it. The group operates on a shared vision and does not need constant direction from the leader. Autonomy is high and the group can, and does, stretch for higher-level goals. Disagreements and issues are quickly addressed and solved. The group is self-sufficient, only needing the leader for clarification or for mentoring. Questions can be answered by any group member because they are all in tune with each other.

A fifth stage was added in 1977, adjourning, which is about dissolving the group, which if unplanned is very stressful. This stage features the completion of tasks, termination of roles, and reduction of dependency. Some refer to this stage as "mourning" because it may even affect former participants (Smith 2005). Adjourning is breaking up the group when tasks are completed, purposes are met, and new challenges are available. Members are now closely bonded and relationships exist, so leaders must attend to emotions, vulnerabilities, and sensitivities. Sympathy and empathy from the leader, and other group members, is very important. Questions are typically about what happens next to the team and to their shared accomplishments.

This is all great information about groups. But as we focus back on interpersonal communication for individuals, we must ask ourselves why we communicate. Here's a short list that I'm sure anyone can make longer.

- Prove a point
- Determine relationship status
- Self-defense
- Persuasion
- Understanding
- Pursuit of power
- Unite a group
- Elicit a reaction from another

Once the group is together, cohesiveness is important. Cohesiveness is a force that influences members of the group to remain and participate. It can also be the amount of attraction that members of the group feel toward other members of the group. There are two types of group cohesion. Task cohesion addresses the degree to which members are working toward a common goal. Social cohesion estimates the level of good feelings internally, or positive affect, among group members. Communicators should recognize all of the variables that might have an effect on members' attraction to a group:

1. Potential to gain security, affiliation and recognition
2. Ability to gain resources and social standing through participation
3. Understanding of the pros and cons of group membership
4. Comparing the current group to past experiences

Group dynamics are important, but so is the environment in which the group operates. Communication networks should be set up with conditions that are best for collaboration. Communication climate determines the social and psychological tone of the interactive relationships, and addresses the relative acceptance or rejection that group members feel (Muchinsky 1977; Guzley 1992; Atkinson & Frechette 2009). In other words, is the environment conducive to discussion, feedback, and decision? Is the environment supportive or defensive? There are a number of versions of organization climate, but we will focus on the six below:

1. **Clarity:** The understanding of group members about goals, policies, and the requirements of the communication activity. The leader needs to ensure that group members comprehend the benefits and purpose of the communication.
2. **Commitment:** Individual buy-in for a common purpose and achievement of the communication goal or goals. Leaders can begin meetings with motivational moments, reviewing successes and stressing shared achievements and rewards.
3. **Standards:** Setting performance standards and determining the amount of pressure needed to enhance acceptance of the communication goal. The group should set high performance standards and the leader must hold everyone accountable for progress.
4. **Responsibility:** Determine whether people feel personally responsible for success in communication. Delegate responsibility where you can to give everyone a sense of ownership and to make them accountable for the group's success.
5. **Recognition:** Create an atmosphere where people are recognized for their accomplishments and where feedback is the norm. It is important to have a fair balance with the group, applauding superior performance and privately correcting inappropriate behavior or lack of accomplishment.
6. **Teamwork:** Get everyone involved to satisfy belonging needs through cohesion, mutual support, trust, and pride. Share great work across groups and conduct benchmarking to capture the "best of the best."

Armed with this information, we must determine how to create the environment for effective communication. The best way to look at communication in groups and work toward effectiveness is by looking at channels that can affect success and relationship building using Figure 3. Group leaders can build networks that transmit in downward, upward, horizontal, diagonal, or through the grapevine methods (Lunenburg 2011). The networks function to

Figure 3. Communicating in groups

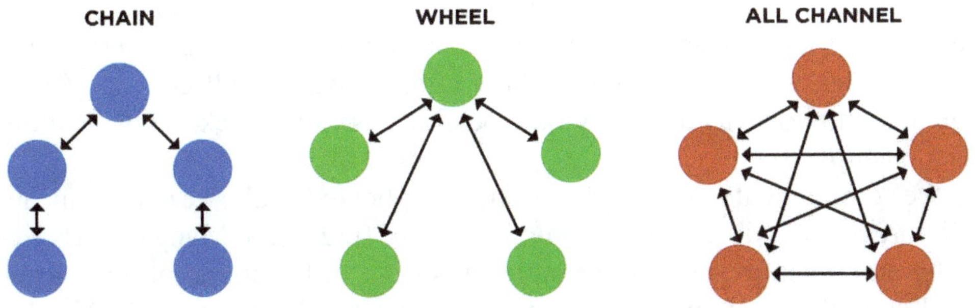

help group members grow, decide, and build relationships. Three communication networks are used to demonstrate the learning on group dynamics: chain, wheel, and all-channel. The chain formally follows the chain of command and the information is usually provided in a relay mode.

The wheel network relies on the leader to act as a central conduit for all of the group's communication. The all-channel network permits all group members to communicate actively with each other. People can communicate freely and build the relationships they value.

It is important to know group dynamics, which some might call structure. To provide more knowledge in this area, we use Structuration Theory. An individual's autonomy is influenced by structure, and structures are maintained and adapted as we communicate. When actor (sender) meets structure, it is called "structuration." (Giddens 1984)

This theory of human behavior is about the duality of structure, which is the synthesis of structure and agency effects. Human behavior is not constrained by educational, religious, or political institutions (structures), for instance. It is not constrained by agency (individual expression of will). Structuration Theory states that human behavior is derived from the interaction of standards, power, meaning, and values, and that there is a dynamic relationship between different aspects of society.

How Do We Communicate?

Now that we have this broad theoretical understanding of interpersonal communication, it is valuable to take a broad view of communication. Why, and how, do we communicate with each other? Most people are goal oriented, doing what they do to achieve something they want or value. Communication is no different. Humans have four basic needs: physical, identity, social, and practical goals. Our daily activities such as eating, drinking, walking, running, or jumping address our physical needs. Interaction with others gives us our sense of identity as we learn how to be "human" from our interactions. Sending and receiving messages allow us to shape our self-concept. Communication allows us to satisfy our social needs as we seek happy, healthy relationships.

We communicate to satisfy basic and complicated goals, we communicate to improve our skills, and we communicate to find a job. Many experts say that communications skills are more important in finding a job than technical competence, work experience, or a specific degree earned. We also communicate to satisfy our needs. Many of us are familiar with Abraham Maslow's Hierarchy of Needs, which addresses physical, safety, esteem, and self-actualization needs. This hierarchy suggests that people are motivated to fulfill basic needs before moving on to other, more advanced needs (Maslow 1970). The hierarchy is normally displayed as a pyramid, where the lowest levels are made up of the most basic needs. These are basic physical requirements, like food, water, sleep, and warmth, and once they have been met people move on to the next level of needs. Safety and security are at the next level, and the most complex needs are at the top of the pyramid. As you progress up the pyramid, the needs that are addressed are more psychological and social. The need for love, friendship, and intimacy follow, and then the need for personal esteem and feelings of accomplishment take over.

When we communicate, we try to do it well. We use what's called cognitive complexity to allow us to develop multiple scenarios for viewing an issue, and then find the best way to address it. We use empathy to experience another's situation and develop the relationship. Self-monitoring is very important, because senders should pay close attention to their own behavior if they want to influence the behaviors of others. Commitment is also important to communication, because those who are care tend to communicate more effectively. That might be because they care enough to listen and then give feedback.

We will continue to talk about the importance of feedback during our learning adventure. But there's much more at stake in great communication. There are communication principles that must be addressed to facilitate engaged interaction. This concept will be covered in depth in Chapter 8. In the next section, we will look at similarities and differences in how the theoretical foundations can be applied in FTF versus online communication.

REFERENCES

Afifi, W. A., & Weiner, J. L. (2004). Toward a Theory of Motivated Information Management. *Communication Theory*, *14*(2), 167–190. doi:10.1111/j.1468-2885.2004.tb00310.x

Atkinson, T., & Frechette, H. (2009). *Creating a Positive Organizational Climate in a Negative Economic One: Improving Organizational Climate to Transform Performance*. Retrieved from Http://Www.Trainingindustry.Com/Media/2505214/Creatingpositiveorgclimate_Us_Aug09.Pdf

Barbee, A. P., & Cunningham, M. R. (1995). An experimental approach to social support communications: Interactive coping in close relationships. *Annals of the International Communication Association*, *18*(1), 381–413. doi:10.1080/23808985.1995.11678921

Berger, C. R., & Calabrese, R. J. (1975). Some explorations in initial interaction and beyond: Toward a developmental theory of interpersonal communication. *Human Communication Research*, *1*(2), 99–112. doi:10.1111/j.1468-2958.1975.tb00258.x

Buller, D. B., & Burgoon, J. K. (1996). Interpersonal Deception Theory. *Communication Theory*, *6*(3), 203–242. doi:10.1111/j.1468-2885.1996.tb00127.x

Bullis, C., & Horn, C. (1995). Get a little closer: Further examination of nonverbal comforting strategies. *Communication Reports*, *8*(1), 10–17. doi:10.1080/08934219509367602

Burleson, B. R. (2009). Understanding the outcomes of supportive communication: A dual-process approach. *Journal of Social and Personal Relationships*, *26*(1), 21–38. doi:10.1177/0265407509105519

Burleson, B. R., & MacGeorge, E. L. (2002). *Supportive communication. In Handbook of interpersonal communication*. Thousand Oaks, CA: Sage.

Chatman, J. A., Polzer, J. T., Barsade, S. G., & Neale, M. A. (1998). Being different yet feeling similar: The influence of demographic composition and organizational culture on work processes and outcomes. *Administrative Science Quarterly*, *43*(4), 749–780. doi:10.2307/2393615

Dolin, D. J., & Booth-Butterfield, M. (1993). Reach out and touch someone: Analysis of nonverbal comforting responses. *Communication Quarterly*, *41*(4), 383–393. doi:10.1080/01463379309369899

Feng, B. (2009). Testing an integrated model of advice giving in supportive interactions. *Human Communication Research*, *35*(1), 115–129. doi:10.1111/j.1468-2958.2008.01340.x

Fisher, B. A., & Ellis, D. G. (1980). *Small group decision making: Communication and the group process*. McGraw-Hill.

Floyd, K. (2015). *Affection Exchange Theory. In The International Encyclopedia of Interpersonal Communication*. John Wiley & Sons, Inc.

Giddens, A. (1984). *The constitution of society: Outline of the theory of structuration*. Univ of California Press.

Guzley, R. M. (1992). Organizational climate and communication climate predictors of commitment to the organization. *Management Communication Quarterly*, *5*(4), 379–402. doi:10.1177/0893318992005004001

Hofstede, G. (1991). *Cultures and Organizations: Software of the Mind*. London: McGraw-Hill U.K.

Jacobson, D. E. (1986). Types and timing of social support. *Journal of Health and Social Behavior*, *27*(3), 250–264. doi:10.2307/2136745 PMID:3772062

Kanter, R. M. (2000). When a thousand flowers bloom: Structural, collective, and social conditions for innovation in organization. In *Entrepreneurship: The social science view* (pp. 167-210). Academic Press.

Knapp, M. L. (1983). *Interpersonal communication and human relationships*. Boston: Allyn and Bacon.

Locke, E. A., Tirnauer, Roberson, Goldman, Latham, & Weldon. (2001). *The importance of the individual in an age of groupism*. Academic Press.

Lunenburg, F. C. (2011). Network Patterns and Analysis: Underused Sources to Improve Communication Effectiveness. *National Forum of Educational Administration and Supervision Journal*.

Maslow, A. H. (1970). *Motivation and personality*. New York: Harper & Row.

Matveev, A. V., & Nelson, P. E. (2004). Cross Cultural Communication Competence and Multicultural Team Performance: Perceptions of American and Russian Managers. *International Journal of Cross Cultural Management*, *4*(2), 253–270. doi:10.1177/1470595804044752

Muchinsky, P. M. (1977). Organizational communication: Relationships to organizational climate and job satisfaction. *Academy of Management Journal, 20*(4), 592–607. doi:10.2307/255359

Neff, L. A., & Karney, B. R. (2005). Gender differences in social support: A question of skill or responsiveness? *Journal of Personality and Social Psychology, 88*(1), 79–90. doi:10.1037/0022-3514.88.1.79 PMID:15631576

Petronio, S. (2015). *Communication Privacy Management Theory. In The International Encyclopedia of Interpersonal Communication*. John Wiley & Sons, Inc.

Smith, M. K. (2005). Forming, Storming, Norming and Performing in Groups. In The Encyclopedia of Informal Education. Academic Press.

Tuckman, B. W., & Jensen, M. A. C. (1977). Stages of Small-Group Development Revisited. *Group & Organization Management, 2*(4), 419–427. doi:10.1177/105960117700200404

Wagner, J. A. (1995). Studies of individualism-collectivism: Effects on cooperation in groups. *Academy of Management Journal, 38*(1), 152–173. doi:10.2307/256731

Wall, B. (2008). *Working relationships using emotional intelligence to enhance your effectiveness with others*. Academic Press.

Weeks, H. (2008). *Failure to communicate: how conversations go wrong and what you can do to right them*. Boston, MA: Harvard Business Press.

Section 3
Engaged Interaction

Chapter 8
Where Do We Go from Here?

OVERVIEW

We have covered a lot of ground up to this point. Now, we have a chance to look back at what we have learned, and to discuss how we will use the knowledge to move through the rest of our communication journey. The first chapter started out talking about the challenge involved with moving from strictly online interactions to face-to-face ones. The challenge, of course, is about using the full range of the communication process from sender to receiver to feedback. The feedback cycle was covered, followed by information about communication barriers and how to overcome them. We focused on an understanding of the characteristics of social networks and interpersonal communication and how they relate to each other. That chapter also covered the importance of a mutual agreement between the parties, the need to read social cues, and the importance of full-range communication from sender to receiver to feedback.

Chapter 2 covered how technological advances caused an explosion of social media and digital communication. This is an analysis of how people use technology to interact with others, which is called computer mediated communication. CMC is not bound by the limitations of geographical distance and time. The technological advances have unbelievably improved the smartphone, which is now a mega-functional, somewhat pocket-sized computer that includes a mobile app for virtually all of the capabilities that we used to get only from our desktop or laptop computer. This chapter shares information from an IT intelligence report that states that half of the world's

DOI: 10.4018/978-1-5225-1897-6.ch008

population now has a mobile subscription—up from just one in five 10 years ago, and an additional one billion subscribers are predicted by 2020, taking the global penetration rate to approximately 60%. That chapter is focused on getting online communicators to see, understand, and accept the skills and benefits of face-to-face communication, so the rest of the chapter covers the motivation, cultural considerations, and issues of multiethnicity.

Chapter 3 discussed communication theories relevant to social networking to help with communication knowledge and discovery. The theories presented in this book are focused on enhancing participation by addressing the necessary improvements to identify determinants and opportunities that affect levels of acceptance and participation. The emphasis in examining social networking should be on creating value that improves participation. The discussion focused, in part, on presenting value propositions that promote and enhance participation. This was followed by a focus on the communication gap between FTF and online, using emotional intelligence, sensemaking, and communication accommodation theories (CAT). There is an identified need to focus on important interaction considerations, like language, context, identity, and intergroup and interpersonal factors.

Chapter 4 featured a high-touch approach to social capital to identify ways to improve and enhance social networking participation. Readers were introduced to Social Networking and Individual Performance, or SNIP, created to pursue success in digital communication with an emphasis on reinforcing organizational training and improving internal and external communications through social media. The chapter focused on energizing the use of social capital with considerations for building trust and seeking engagement. The action plan for success provided a tool that can build and strengthen social capital for people and organizations.

Chapter 5 covered communication methods and compared barriers to communication between FTF and online interactions. It provided information and illustrations about full-range communication using knowledge from Michael Argyle's 1972 Communication Cycle, Wilbur Schramm's 1954 Model of Communication, and the Shannon-Weaver Model of Communication. The Shannon-Weaver model demonstrates communication as an interactive process where participants exchange messages and deal with noise in the channel. The chapter introduces several important references for interpersonal communication discovery and discussion. These references provide a foundation for continued communication discovery. The chapter concludes with a closer examination of barriers to communication.

Chapter 6 discussed the gap in terms of what method works best in addressing it, demonstrating a high touch effort to get all the knowledge in one place. It reviewed the three major theories that provide the foundation for analysis: EI, sensemaking, and CAT. These theories do not, however, take into account a more detailed examination of interpersonal communication that is in Chapter 6. Instead, Chapter 6 works to promote understanding of how to apply each theory and then to understand the relative importance of each method of communication. This knowledge is intended to allow communication participants to interact efficiently between the two methods regardless of the direction of movement. This examination is a way to keep senders and receivers in tune, allowing full human contact at all times. The chapter is also intended to help online communicators discover a level of comfort away from the digital conversation so that they can employ verbal and non-verbal cues to interactions, or so that they can find some measure of those cues in the online environment.

Chapter 7 really energized the pursuit of bridging the gap in communication between FTF and online. It provided a focus on interpersonal and group communication, highlighting theories that are relevant for our discussion. Uncertainty Reduction Theory focuses on motivating people to communicate with a goal of reducing uncertainty. Supportive Communication Theory features senders seeking to use verbal and non-verbal behaviors to provide assistance to those in need of improved communication or other services. While its use may be more insightful for online communicators than FTF, Interpersonal Deception Theory states in part that senders try to manipulate messages to be untruthful, and this causes sender apprehension because they worry about getting caught in the false information. Motivated Information Theory combines a three-phase process – interpretation, evaluation, and decision stages – that individuals use to seek or avoid information about an issue. Group behavior theory is also provided, beginning with an analysis of individualistic and collectivistic mindsets. This is an examination of Decision Emergence Theory, the theory of Forming, Storming, Norming and Performing, and Structuration Theory. All of these theories focus on how groups interact and manage conflict.

Communication Privacy Management Theory states that that people have private information, private boundaries, control and ownership about information, and they use rule-based management and privacy management dialectics. Adaptive behaviors that people use in creating and maintaining romantic relationships are addressed using Affection Exchange Theory. After the discovery of interpersonal communication theory, the chapter addresses these activities as interdependent arrangements where feedback is crucial to success. There is also a discussion of the importance of conflict resolution, where interaction participants work to save face, correct the record, determine winners or losers, or to set terms for future interactions.

We finish this chapter with interesting tools to improve the communication activity. We start with the Social Networking Matrix. The chart provides the theory that is covered earlier in the book followed by information on the basics of what the theory describes. The matrix also provides two suggested approaches each for FTF and online interactions. Following that, the Interpersonal Communication Matrix covers interpersonal communication theories. The matrix offers reasons for interactions that are based on the theory, followed by two suggested approaches for FTF and online. Finally, Engaged Interaction is introduced as a resource that fuels use of the triad of listen, hear, and understand. Engaged Interaction features sender and receiver fully committing to the communication activity to achieve the triad.

Application of the Knowledge

What do we do with the knowledge we've gained? We need a plan of action, one that brings all of our research and knowledge together so we can communicate in a deliberate, efficient manner. To craft the plan, we need to really understand what we've experienced in the preceding chapters of this book. There is a need to conduct full-range interactions to bridge the gap in communications between FTF and online activities. For various reasons, social media communicators don't have access to verbal and visual cues that can make interactions easier. When social media communicators have access to the human touch in terms of verbal and visual cues, they may not take advantage of them. It has been well documented to this point in the book that verbal and non-verbal information can make the communication rich in information sharing and relationship building. These "gifts" are important because communication participants should create and strengthen bonds that turn into open deliberations and, hopefully, relationships.

We have examined numerous social networking and communication theories that provide valuable information contributing to communication success. We owe it to ourselves to communicate in the best way possible every time, and that requires full use of all available tools. This is a high-teach moment. This book is arming every reader with the resources to be better communicators. This is a call to action to make use of the information. As you move through the rest of this learning journey, you will find resources that will be valuable in any communication event. As you continue on this learning journey, take advantage of the resources provided. Face-to-face communicators have great opportunity to use the human touch to bridge the communication gap by taking advantage of verbal and non-verbal cues. However, there are ways that online communicators can get closer to full-range communication as well. Our learning from this point on will continue to highlight the strengths and weaknesses of FTF and online communication.

Let's experience some of the resources mentioned above. Two charts are provided to help achieve full-range communication. The Social Networking Matrix (Table 1) and the New Age Communication Matrix (Table 2) appear in the pages that follow with explanations for their use.

Table 1. Social Networking Matrix

Social Networking	The Basics	FTF	Online/Digital
		Two Approaches	
Innovation Adoption	Positive connection based on goals.	• Create positive atmosphere • Be supportive of receiver	• Create positive atmosphere • Provide feedback link
Diffusion of Innovations	Connect via structure and demographics.	• Focus on social structure • Find common ground	• * Analyze your audience • * Find common ground
Social Capital	**Create value through relationships.**	**• Build relationships • Seek trust & belonging**	**• * Build relationships • * Seek trust & belonging**
Cognitive Dissonance	Personal inconsistency causes dissonance, driving need for resolution.	• Change beliefs & actions • Change action perception	• * Change beliefs & actions • * Change action perception
Knowledge Gap	Greater access to info creates greater ability to acquire more.	• Increase info sharing • Solicit questions	• Enhance info sharing • Emphasize group chat
Social Presence	**Do we perceive the person as real?**	**• Build relationships • Seek trust & belonging**	**• Ask questions • Seek verification**
Social Network Analysis	Identify levels of participation.	• Increase info sharing • Solicit questions	• Build strong ties • Improve structure/ practices
Social Cognitive	Human action results from environment, personal trait and behavior interaction.	• Revisit prior successes • Set conduct standards	• Model & imitate others • Set conduct standards
Social Learning	**Continuous interaction between cognitive, behavioral and environmental influences.**	**• Tailor message to receiver • Motivate the receiver**	**• Tailor message to audience • Motivate your audience**

NOTE: The approaches marked with * can be difficult to impossible unless the sender can verify that the receiver is a "real" person

The matrix on the previous page covers various social networking theories. This is a listing of relevant theories and the basics of what each theory is about. The matrix guides the reader through suggested approaches based on the communication method: FTF or online interaction. For instance, diffusion of innovations theory discusses connecting via structure and demographics. In using this theory for communication, face-to-face senders should focus on social structure and find common ground in interactions. The common ground refers to matching demographics of sender and receiver where possible. The online sender should analyze the audience to achieve success and, like the FTF sender, should seek common ground.

It is important to add one note of caution or clarification to the matrix. There are three items in the matrix marked with an asterisk (*) in which the sender has to be careful. For these items, approach can be difficult to impossible unless the sender can verify that the receiver is a "real" person. In social presence theory, we are cautioned to determine whether the receiver is real, which is a challenge in online communication. If the receiver is an eggshell (just a silhouette), or a picture of a cat, or some object, the sender must determine whether that is a person they can connect with. It is very challenging to take the actions mentioned – audience analysis, finding common ground, building relationships, seeking trust, changing beliefs, and changing perceptions – unless you can determine the receiver's realness. Communicators need to know who they are speaking to and determine the person's interest and values before they can work on shared value in the communication. This is important information, so the message can be tailored to the audience and we can determine at some level whether or not the person will be interested.

There are three items in this matrix that should be used most often when communicating or attempting to improve communications. First, social capital is absolutely crucial to any human interaction, whether it is in person or online. Both FTF and online communication need the sender to focus on building relationships and on seeking trust and belonging to connect with the receiver. Social capital is valuable because it's classifications into types provide us with excellent ways to use social capital, and the network and structural types can be used interchangeably. It has a network perspective and a structural perspective, allowing us to view communication from the network or personal perspective, or from the structural or organizational perspective.

Social capital can be classified as bonding, bridging, or linking from the network perspective. Bonding is defined by ties between individuals who have a high degree of connection, like those within the same social group, or horizontal ties. Bridging social capital is ties between individuals who are in different social groups, or vertical ties. These ties are often characterized by a lack of formal hierarchical structure; they have a moderate level of trust that allows communication and/or access and benefits. Linking social capital is characterized by explicit, formal, or institutionalized power or authority, where norms of respect and trust allow interactions across networks. In fact, bridging and linking social capital are very similar, primarily differing in terms of the level of trust: moderate or low for bridging, higher for linking.

It is also important to look at social capital based on the source or structural perspective, which would be structural, cognitive, or relational. This type is about established roles and social networks that are supplemented procedures, precedents, and rules. Social transactions need this structure and stability. The structure and stability are built from culture and institutional foundations. Cognitive social capital is about shared norms, values, attitudes, and beliefs. This type moves people toward collective action that is mutually beneficial, which is very important for communication activities of any type. Social relationships that are built primarily on trust and trustworthiness foster relational social capital. This is really about the connection that individuals have and the character and conduct of that connection. Some examples should make social capital more clear. The ties that gangs create illustrate bonding social capital, while the interactions of sports teams or music groups, for instance, demonstrate bridging social capital. A person's ongoing communication with a religious leader or a mentor would illustrate relational social capital. All of this social capital knowledge is important because communication is needed if we are to access and use social capital to identify problems and their solutions, exchange information, and manage conflict. Effective communication requires some level of sharing context between sender and receiver, and the cognitive dimension of social capital is relevant. This dimension highlights the important shared meaning and understanding between communicators.

The second high-use theory is social presence. This theory addresses whether the sender perceives the receiver as a real person. In FTF, of course we know the person is real, but we must be concerned with a different kind of reality. This kind of reality is what was covered when Chapter 1 discussed perceived improvement potential (PIP), which allows us to make a conscious determination about whether this activity will bring personal benefits (Brown 2011). Senders want to communicate in an environment that builds trust, fosters interactions, and addresses shared needs with the receiver. PIP seeks reality in the individual in terms of their intention to behave in a certain way, and helps the sender determine whether the receiver will actually exhibit the desired behavior. Reality for the online sender is all about figuring out whether the person is who they say they are. The desired approach is to ask questions and seek verification of information to ensure we are communicating with a "real" person.

The third highly relevant theory is social learning. Social learning deals with continuous interaction between cognitive, behavioral, and environmental influences in the communication. FTF and online senders should always tailor the message to the receiver, and make sure the receiver is motivated to enter and continue the conversation. FTF senders can use the visual and verbal cues in the environment to attend to the influences that are important in social learning. Online senders need to be more creative to adjust to cognitive and behavioral influences, but do have a limited need to pay attention to environmental influences.

Face-to-face senders have a range of actions they can take to address the other theories in the matrix. These actions may work across theories if senders decide to interweave them to get the best outcome for message delivery. The actions include creating a positive atmosphere, being supportive of the receiver, focusing on social structure, finding common ground, changing beliefs and actions, changing perceptions of the action, increasing information sharing, soliciting questions, revisiting prior successes, and setting standards of conduct. Effective communicators will be skilled at employing a variety of approaches to get the best of the interaction.

Online communicators can also work across theories. Their interactions may include some or all of these actions: create a positive atmosphere, provide a feedback link or outlet, analyze the audience, find common ground, change beliefs and actions, change perceptions of the action, enhance information sharing, emphasize group chat activities, build strong ties with receivers, improve communication structure and/or practices, model and imitate effective practices of others, and set conduct standards.

This is a high-teach moment because this is where the gap in communication is evident. In FTF interaction, you can look a person in the eye and figure out whether or not they are real. They may be real yet untruthful, but FTF communicators can ask questions and use verbal and non-verbal cues to get the true picture. This ability does not fully exist in online communication. That means that there is a potential blind spot in the interaction, and that can be very damaging to full-range communication. There is a note on the matrix telling the reader about potential limitations based on the online environment. If the online sender is not sure whether the receiver(s) is a real person, it is hard to verify the effectiveness of the communication.

Senders must master interpersonal communication, which is as important as using social networking theory. The Interpersonal Communication Matrix (Table 2) provides a tool for success.

Table 2. Interpersonal Communication Matrix

Online Communication Theories	Interact to:	Two Approaches	
Emotional Intelligence	**Recognize and employ one's emotions**	**• Ensure shared value • Seek and give feedback**	**• * Ensure shared value • * Seek and give feedback**
Sensemaking	**Create meaning from social context**	**• Share responsibility • Adjust to barriers**	**• Seek info sharing buy-in • Adjust to barriers**
Communication Accommodation	**Manage personal and social identity**	**• Agree on identity • Adjust to receiver reaction**	**• Create social identity • * Seek and give feedback**
Uncertainty Reduction	Alleviate uncertainty	• Develop relationships • Assess verbal & non-verbal	• * Develop relationships • * Seek and give feedback
Supportive Communication	Provide needed assistance	• Send supportive messages • Provide assistance	• Send supportive messages • Provide assistance
Interpersonal Deception	Some communicators will be untruthful	• Tell the truth • Seek more information	• Tell the truth • Seek more information
Motivated Information	Communicators seek or avoid the issue	• Foster open dialogue • Focus on the issue	• Foster open dialogue • Focus on the issue
Communication Privacy Management	Protect privacy and control information	• Determine what to share • Set rules for sharing	• Determine what to share • Set rules for sharing
Affection Exchange	Achieve long-term relationship	• Develop relationships • Assess verbal & non-verbal	• * Develop relationships • * Ensure receiver is "real"

NOTE: The approaches marked with * can be difficult to impossible unless the sender can verify that the receiver is a "real" person

The matrix on the previous page covers interpersonal communication theories, and then provides the reason for interaction based on the theory, followed by two suggested approaches for FTF and online. A good example in this chart is the last entry, affection exchange theory, which is about achieving long-term relationships. FTF senders should focus on developing the relationship and on assessing verbal and non-verbal cues. The online sender should focus on relationship development, but should also ensure the receiver is real, or in other words, that they are who or what they say they are.

The matrix begins with the three theories that are the foundation of our analysis of the communication gap that exists between FTF and online communication. They represent the key to good interpersonal communication. These were introduced in Chapter 5 to demonstrate the methodology needed for the high-touch effort that puts all of the knowledge in one place. EI, sensemaking, and CAT allow our discovery and promote understanding of ways to apply the theories and illustrate the relative importance of each method of communication that we examine. Senders and receivers can use this knowledge to interact efficiently in FTF or online. Most importantly, it provides skills that online communicators can use to introduce or improve application of the human element in interactions. That is progress. Now, the overview of theory and action with concrete suggestions allows FTF communicators who move to online interactions to envision the best approach for their messaging. The two suggested approaches are the best ways to get started. Online communicators would do the same, using the FTF approaches to their best advantage.

We know that EI is about recognizing and employing one's emotions in communicating (for our purposes). Self-awareness helps us develop people skills to listen, hear, and understand, and to take other's emotions into account. FTF and online communicators should ensure shared meaning and give feedback to ensure value in the interaction and to achieve the best outcome. In sensemaking, senders work to decode messages using a common language to drive our actions in the right direction. FTF senders are urged to share responsibility and adjust to barriers as they create meaning from the social context. Online communicators should seek buy-in for the information sharing and adjust to barriers that might make the communication difficult. CAT is focused on managing social and personal identity with an emphasis on highlighting or minimizing the differences between sender and receiver during messaging. Language, context, identity, and intergroup and interpersonal factors are vital to good communication. FTF senders need to agree on identity with receivers, and then adjust to receiver reactions. For online communicators, the emphasis is on creating social identity and seeking and giving feedback.

Proper use of the Interpersonal Communication Matrix is the same as the Social Networking Matrix; senders can work across the theories for the best results. There are many forces in the communication atmosphere as Figure 1 demonstrates.

FTF senders should focus on these actions across the other theories: develop relationships, assess verbal and non-verbal cues, send supportive messages, provide assistance, tell the truth, seek more information, foster open

Figure 1.

dialogue, focus on the issue, determine what to share, and set rules for information sharing. Online communications can focus on these actions: develop relationships, seek and give feedback, send supportive messages, provide assistance, tell the truth, seek more information, foster open dialogue, focus on the issue, determine what to share, set rules for the information sharing, and ensure the receiver is "real."

The explanation of the Interpersonal Communication Matrix is important because it gives us important information about bridging the communication gap between face-to-face and online interactions. The matrix lists 18 actions that senders can take to apply theory to their communications, and 13 of the suggested actions are the same as listed in Table 3.

Table 3. Conflict style adjustments

Adjusted Conflict Styles
1. Ensure shared value
2. Seek and give feedback
3. Adjust to barriers
4. Develop relationships
5. Send supportive messages
6. Provide assistance
7. Tell the truth
8. Seek more information
9. Foster open dialogue
10. Focus on the issue
11. Determine what to share
12. Set rules for sharing
13. Develop relationships

That indicates that we have the means to bridge the communication gap. What we have to do is come up with a systematic, reliable way to look at the gap, understand the gap, and bridge the gap. We do that by focusing on the triad mentioned in Chapter 6: listen, hear, and understand. We do that through engaged interaction, which was briefly introduced in Chapter 6.

Engaged Interaction

Engaged interaction shown in Figure 2 is accomplished by senders who employ full-range communications with a specific goal: that both parties will listen, hear, and understand, and that the communication will continue until that happens.

Engaged interaction is defined as employing flexible, full-range communications to ensure that both parties *listen, hear, and understand*, and continue the interaction until that happens. Engaged interaction starts with *listening*. Mark Twain said, "If we were meant to talk more than listen, we would have two mouths and one ear." Great communicators know when to stop talking and listen. They try to relax so they are receptive to what is being said. Senders try to put the receiver at ease so they are comfortable engaging and providing feedback. At every stage of the communication, it's important to focus on what is being said by removing any distractions, whether or not they are communication barriers. Try to understand the other person's point of view and be patient to allow the interaction to work. Good communicators are impartial and they listen for ideas, not just words. They understand that communication is a cycle, and encoding, decoding, and feedback are equally important to the success of the interaction. Always remember to use non-verbal – gestures, facial expressions, eye movements – to get the most out of the communication.

Figure 2.

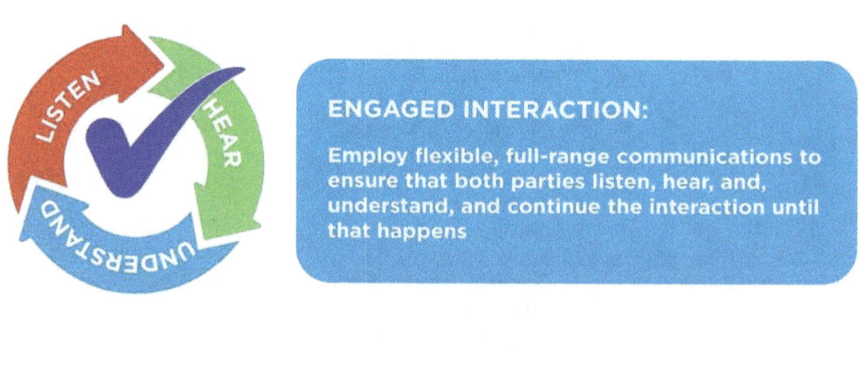

Senders need to be skilled in *hearing* to be available for the response at any time. Avoid multitasking so that you can hear exactly what the receiver is saying. Make eye contact and block out distractions so you can focus on the shared information. Senders should ask questions and restate what they hear to verify that the message has been sent and received. This gets the sender ready for feedback and continued communication.

Empathy is important to allow senders to *understand* throughout the communication process. Empathy means seeing what others see and understanding what they are feeling about what they see. Empathetic senders try to understand why the other person's actions make sense. There are three parts to empathy: understanding yourself, understanding others, and non-verbal empathy. Know yourself by accepting your feelings and emotions to engage open and honest interactions. Know others so you can understand their thoughts, feelings, and commitment to the communication activity. Once you understand others, you can tailor your non-verbal communication to best suit their needs.

Beyond the broad understanding above, engaged interaction has several important parts that require understanding and execution. The parts are:

1. Communication Principles
2. 4Rs of New Age Communications
3. Strategic Flexibility

Communication Principles

Use of communication principles requires that we acknowledge their existence and understand their usefulness. We use various channels to conduct two-way communication. This is a complex process, and we have to use our own filters to determine what we say and hear. Senders should acknowledge the various aspects of communication that can be used to promote success, like sending concise messages, being considerate in terms of crafting the message with the receiver in mind, being courteous to enhance bonding with the receiver, and being accurate with the message. These factors are the glue that completes the full cycle of encoding, delivery, decoding, and feedback, and senders should understand their use.

Let's talk about understanding. Communicators should realize that not all communication seeks understanding, and more communication does not necessarily improve things. In fact, communication will not solve all of your problems. Let's face it, effective communication is not a natural ability, but we can all learn to do it well. Engaged interaction is the way to "listen, hear and understand," and that triad leads to relationship building and trust. Effective communication is about goal achievement through actions that create or improve the relationships where they occur.

How do we really execute engaged interactions? We do it by focusing on audience, topic, and feedback and using the tools listed in Table 4 to promote good interactions. Know your audience and work to determine the best way to get the audience interested in the message. Using multiple communication techniques will allow success with a variety of receivers in a variety of situations. When delivering the message, have a clear understanding of your purpose for addressing the topic. Anticipating objections, presenting a balanced view of the topic, establishing credibility, and following through on actions is the best way to avoid getting sidetracked during message delivery. Communicate in short bursts and change the way the information is presented when necessary. Senders should develop a useful, practical way to get and give feedback. All of these activities contribute to engaged interaction. Experts say that people remember 10% of what they read, 20% of what they hear, 30% of what they see, and 40% of what they hear and see. In other words, as engagement in the triad (listen, hear, understand) increases, so does the value of the communication activity. Here is a short list of actions that promote engaged interaction.

Good communication practices also contribute to engaged interaction. Address barriers to communication by managing stress and out-of-control emotions. Work to stay focused on the message, and don't be guilty of inconsistent or negative body language. Don't say you're interested in what the person is telling you while you're looking out of the window, and don't say you have time and then keep looking down at your watch. Become an engaged listener who focuses not just on what the person is saying, but on how they say it and how they are feeling. This gets you ready to give feedback, empathy, or support as necessary. Keep your focus on the speaker, avoid interrupting the flow of the message, be truly interested in the conversation, and don't

Table 4. Tips for engaged interaction

Promoting Engaged Interaction
1. Listen
2. Improve listening skills
3. Listen to the complete message
4. Eliminate distractions
5. Focus on the content of the message, not just the form
6. Weigh emotionally-charged language
7. Think critically and efficiently when communicating

judge, just analyze and respond. Pay attention to non-verbal signals and individual differences. Match non-verbal signals with your words, let your body language deliver positive feedback and input, and adjust your non-verbal signals to be appropriate for the interaction. This takes thought and practice. Avoid stress for yourself and the other party. The way to do this is to stall by asking for the question to be repeated or requesting a clarification to give you time to get control. State your position clearly, pause for effect or to think, and summarize what you think you heard. Finally, provide concrete examples for the information you present. You can also agree to disagree, compromise, or find humor in the situation to relieve stress. All of these practices are good for the communication interaction and all parties involved.

So, as we acknowledge and understand communication principles we must keep four things in mind:

1. Communication is inescapable, whether it is intentional or unintentional.
2. Communication is irreversible, because you can't "unsay" what you've said.
3. Communication is contextual, taking into account psychological, relational, situational, environmental, and/or cultural aspects.
4. Communication is complicated, because there are any number of good and bad consequences at play.

4Rs of New Age Communications

A "4Rs" chart is presented in Figure 3. It provides a set of actions that can be used on a recurring basis to achieve success.

Figure 3. 4Rs of New Age Communications

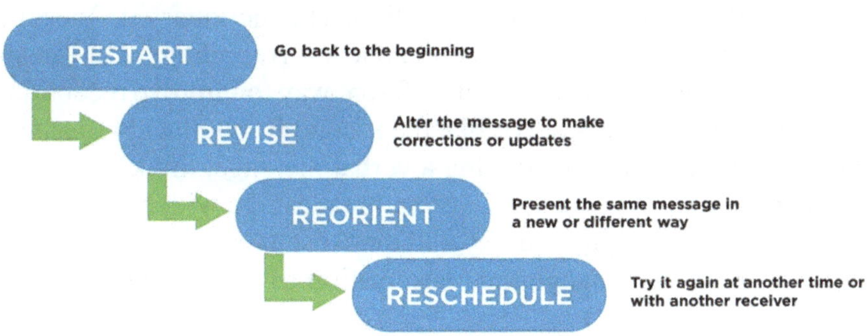

The 4Rs are important for managing communications, or for improving or reshaping difficult interactions: Restart, Revise, Reorient or Reschedule. These are approaches that can be used to energize the communication while it is in progress. The first "R" is restart the conversation. Go back to the beginning of the interaction and get a fresh start, or just rewind a bit and focus your efforts. Restarting is a second chance. It gives both parties an "escape" so that they can change the communication in a positive way. When you restart the conversation, leave the past behind as much as possible. Don't revisit the difficulty unless it's absolutely necessary to chart a new course. Clear the air and start communicating!

The second "R" is revise the conversation. Change the focus, the tone, or the language you're using to get better understanding and seek agreement on your message. This approach is about changing to get to your desired end state. If the sender is not coming to a "meeting of the minds" with the receiver, a great approach is to change the focus of the communication. For instance, if the interaction is about motivation and you're using sports analogies, you might find that the receiver is not well versed in sports, or just isn't interested. You can change the focus to getting ahead in business, winning a performance award, or playing a game of checkers. In terms of tone, if you are communicating and using humor, the receiver might not get the joke, or might not think you're funny. Revise the approach to a more serious tone, possibly just covering factual aspects of the message and its intended outcome. Language does not necessarily refer to speaking the same language as the person. Instead, it's about the words that you use. If the sender is delivering the message and using military acronyms and jargon, for instance, that is a very different language than what the receiver might be used to. If they have never been in the military, they may not comprehend the information that is being exchanged. Senders must ensure that the language being used is common to both parties.

The third "R" is reorient the conversation. The dictionary defines this as finding one's position again in relation to one's surroundings. This means to present the same message in a new or different way to improve reception and enhance decoding. This is kind of like the second R, but the approach really is to use your imagination. If sending a verbal message, think about changing it to a visual one. If you are delivering the message with an instructional video, consider changing to a survey and then discussing the results with the group. Be creative and change the conversation to ensure the message can be decoded by the receiver.

The fourth "R" is reschedule. Try the conversation at another time that is mutually beneficial, or try it with another receiver. You might even try the conversation in another setting. For instance, the receiver might be intimidated or uncomfortable getting the message in your office or your building. Schedule the interaction at a neutral location to put the subject at ease.

Whichever "R" you choose, you should then take 3 important actions to succeed:

1. Be in the room,
2. Avoid distractions, and
3. Make the connection.

Once you've chosen your "R," or a combination of "Rs," you must take action that leads to success. First, be in the room. When you're communicating, give creation of the message, delivery of the message, and decoding of the message your utmost attention. Focus on how the receiver is responding and make the necessary adjustments to keep the conversation alive and encourage feedback. Second, avoid distractions, especially multitasking. People may be good at one thing, or two things, or three things, but very few can handle more than that at one time. And multitasking can be any number of activities that take away from the ability to focus on the receiver and his or her reception and decoding of the message. Third, make a connection. Let the receiver know that they have your attention during the communication interaction. Show that you can give your undivided attention to the receiver and the interaction. Remember, you created the message and decided it was important enough to deliver, so sustain that importance through decoding and feedback activities.

Strategic Flexibility

After considering communication principles and the 4Rs, we turn to strategic flexibility. An examination of strategic flexibility research illustrates its use in organization theory, marketing, economics, and strategic management settings. Our analysis evaluates several definitions of strategic flexibility to determine the one best suited to the pursuit of effective interpersonal communication. To paraphrase several of the previous definitions, strategic flexibility is the ability to take action in response to external environmental changes, or to drive intentional changes and adapt to environmental changes. The overall basis of the theory is that the strategy is flexible to address variations in the issue at hand and that it can be implemented or changed rapidly (Slack & Nigel 1983; Evans 1991; Bahrami 1992; Sanchez 1995; Nadkarni & Narayanan 2004; Roberts & Stockport 2009). For instance, strategic flexibility allows companies to respond quickly to competitive conditions by changing to create or maintain competitive advantage (Hitt et al. 1998, p. 26). One related concept to strategic flexibility is organizational flexibility, which is "… the ability of the organization to adapt to substantial, uncertain, and fast occurring environmental changes that have a meaningful impact on the organization's performance." (Aaker & Mascarenhas 1984)

Another way of looking at strategic flexibility in groups is to note that there is an ability to react to changes in the external environment and then commit resources to affect change with the intent to halt or reverse commitments (Shimizu & Hitt 2004). In other words, groups are flexible in that they can adjust to influences on their operations. They may adjust to internal or external forces, but strategic flexibility is focused on external forces.

While most scholars equate this to group communication, this book uses it in an individual sense. Regardless of whether we think in terms of group or individual communication, the intent is the same. Strategic Flexibility is about expanding the communication toolkit to enable use of the best possible skill or behavior for a particular situation. Strategic flexibility means senders examine the external environment and then use what they learn to determine major changes in the actions they take (Shimizu & Hitt 2004). Once the examination is finished, the sender quickly commits to new courses of action. The key is to recognize the issue and act promptly when there is a need to delay, cancel, or reverse the existing group commitments. *So, strategic flexibility is the act of changing the message in response to internal and external influences to increase the chances that the message will achieve its desired result.*

Further examination of research and social media postings on strategic flexibility provide more data that is helpful. Combining the work of several scholars, we arrive at several steps to strategic flexibility:

1. Assess the situation;
2. Evaluate the communication environment;
3. Use various skills (tone of voice, gestures, expressions, body language, etc.);
4. Be creative;
5. Don't be afraid to adapt or change; and
6. Reassess and reevaluate.

Katsuhiko Shimizu and Michael A. Hitt discussed the insensitivity, self-serving interpretation, and inaction that can plague managers (2004). They recommended six practical ways to avoid these problems, stressing that managers and organizations need to take a proactive stance to overcome biases and to guard against becoming rigid in decisions and actions. The actions are (2004):

1. Measure and monitor decision outcomes.
2. Stimulate decision-making processes by incorporating a devil's advocate approach.
3. Create dynamic mechanisms to gain new ideas and perspectives from outside of the firm, before problems appear.
4. Recognize the limitations of static governance systems.
5. Do not narrowly focus on one decision. Consider decision portfolios.
6. Analyze and measure learning that can be used in the next step.

The authors' primary objectives were "….(1) to help managers understand the importance and difficulty of developing strategic flexibility and (2) to provide practical recommendations to deal with this challenging task by focusing on structural issues (i.e., contingencies) that contribute to various biases and traps." (2004) The current research also suggests that the capability requires focusing the sender's attention, completing an assessment of the situation, and always taking action in response to what we experience. In communicating, leaders must also learn to cope with uncertainty in the environment. For our purposes, we merge all of the suggested actions we have covered into five:

1. Stimulate decision-making with open dialogue;
2. Exploit internal and external relationships;
3. Focus on multiple methods and venues;
4. Reevaluate and reassess your actions; and
5. Make any necessary course corrections.

When we discuss open dialogue in the first step of our strategic flexibility, we are referring to restarting or revising the communication (4Rs). Remove emotions from the communication before the restart or revision. This will insure there won't be any perception of a threat. Take some physical precautions as well. Sit down to communicate because that demonstrates a calm and composed communicator who has a sense of urgency but cares about the person or people at hand. Manage your tone of voice and take regular pauses, resulting in a relaxed delivery and a relaxed leader and/or sender. These are great suggestions to promote calm, ease into (or back into) the conversation, and to help the restart or revision to achieve the desired effect.

Our second step, exploiting relationships, is about reorienting (4Rs) the conversation using social capital or, more simply, by getting in touch with people to energize communication. Establish important connections based on shared value. Messages are received better when we give the person a sense that they are the focus of our communication and we view them as peers, not people we need to educate. We may need to educate people; we just don't want to be condescending about it.

The third step is emphasizing multiple methods and venues for our interactions. This step is really about rescheduling (4Rs) the conversation with the right method in the right venue to ensure the message is received. Also, don't be afraid to wait until another time to communicate. If the message is valuable, it will still be valuable at a later time in most instances. The "right time" is an important factor in making communication work. Use language and metaphors that will have great meaning for your intended audience. Try to link your message with their experience by using sports words and metaphors for sporting groups, using inspiring words and metaphors for family or community groups, and using motivational words and metaphors for business groups.

Reevaluating and reassessing our actions is the fourth step and we are applying revision (4Rs). We must take an honest look at how we communicate and make sure it is working. We determine this through feedback with the receiver, but also by examining any actions or consequences that are the result of our communication activities. If we deliver a message and it angers the receiver, then we know we have missed the mark. That's unless, of course, we wanted anger as a reaction. If our message gets the receiver to take an action, but that action is not the one we wanted or intended, then we know our communication may have been misguided. We must reevaluate and reassess during and after the communication activity.

The fifth step is making any necessary course corrections. A good communicator will make the interaction or collaboration right, whatever it takes. Remember that we want to be flexible in our communications, understanding and responding to verbal and visual cues from the receiver or others in the audience. We should also realize that the necessary action to take may be to do nothing, and give the receiver more time to decode the message. A delayed reaction or response might be the best thing for our intended message and outcome. In fact, our actions could be any of the 4Rs: restart, revise, reorient or reschedule.

High-Touch to High-Teach Moment

At this point we've learned a lot about communication in terms of process, research, and theoretical bases, all leading to an action plan. We covered high-touch in terms of conducting effective two-way communication whether FTF or online. Emphasis was placed on using the full communication cycle of encoding, sender, message, receiver, decoding, and feedback. Sections 1 and 2 were geared towards giving you deep understanding of the two methods of communication and an understanding of the communication gap that exists between the two methods. The comparisons and contrasts that have been pointed out so far are intended to provide theoretical and practical knowledge that will help you to communicate better.

In Section 3, we are working on the third valuable communication ability you were promised at the beginning of the book: application resources to help you "bridge the gap" in communication between FTF and online methods. This is intended to put any sender and receiver, regardless of the method, in the best messaging interactions possible. We've started to create the action plan in this chapter. So let's review.

We started by enhancing your knowledge with the Social Networking Matrix and the Interpersonal Communication Matrix. The chapter has given instruction on how to understand and apply the details provided in each matrix, regardless of the method of communication. From there, the chapter introduced engaged interaction, which is vital to the action plan provided herein. Remember, engaged interaction is defined as employing flexible, full-range communications to ensure that both parties listen, hear, and understand, and continue the interaction until that happens. Engaged interaction has three important parts: communication principles, the 4Rs of New Age Communications, and strategic flexibility, which is basically the ability to adapt communication activities to any internal or external influences for the best results.

This chapter provides great resources for communicating. A search through leadership and management, public administration, social networking, and communication literature yields a wealth of information on effective communication. Building trust, for instance, takes time, but it is vital to success. Several things can contribute to building trust. Ensuring that all communication is honest and open is vital. Senders should speak directly to people, saving electronic communication as a means to back up or reinforce the messages you've already sent. Active listening skills and eye contact keep the receiver engaged and signals the genuine interest of the sender.

In addition to trust, people need to feel that the sender is being open and honest with the information and is not holding anything back. Debate is important in communication because disagreements can lead to conflict, either good or bad. Regardless, ideas need to be presented and discussed so that the team is exposed to opportunities and alternate solutions to problems. As the communication continues, it is important to hold everyone accountable for any plans or decisions. Team members should be completely engaged in overcoming obstacles and ensuring progress. Choose the words you use in your message carefully as you work toward a participative atmosphere. Saying exactly what you intend to say is precision, and sending a message that is clear and easily understood is clarity. Both are important to the overall communication. Finally, focus on results that are accepted by and agreed upon by all team members, allowing all to take credit for collective results.

REFERENCES

Aaker, D. A., & Mascarenhas, B. (1984). The need for strategic flexibility. *The Journal of Business Strategy*, *5*(2), 74–82. doi:10.1108/eb039060

Bahrami, H. (1992). The Emerging Flexible Organization: Perspectives from Silicon Valley. *California Management Review*, *34*(4), 33–52. doi:10.2307/41166702

Brown, M. Sr. (2011). *Social networking and individual performance: Examining predictors of participation*. Old Dominion University.

Evans, J. S. (1991). Strategic flexibility for high technology manoeuvres: A conceptual framework. *Journal of Management Studies*, *28*(1), 69–89. doi:10.1111/j.1467-6486.1991.tb00271.x

Nadkarni, S., & Narayanan, V. (2004). Strategy frames, strategic flexibility and firm performance: The moderating role of industry clockspeed. Academy of Management Proceedings. Academy of Management.

Roberts, N., & Stockport, G. J. (2009). Defining Strategic Flexibility. *Global Journal of Flexible Systems Management*, *10*(1), 27–32.

Sanchez, R. (1995). Strategic flexibility in product competition. *Strategic Management Journal*, *16*, 135.

Shimizu, K., & Hitt, M. A. (2004). Strategic flexibility: Organizational preparedness to reverse ineffective strategic decisions. *The Academy of Management Executive*, *18*(4), 44–59. doi:10.5465/AME.2004.15268683

Slack, N., & Nigel, S. (1983). Flexibility as a Manufacturing Objective. *International Journal of Operations & Production Management*, *3*(3), 4–13. doi:10.1108/eb054696

Chapter 9
Face-to-Face or Online, Online or Face-to-Face?

OVERVIEW

This chapter compares FTF and online communication and suggests the best ways and times to use each method. At times, the approach will seem like a criticism of online communication, but that is not what is being done. No approach is all things to all people, so we have to understand the best path to success. There are positives and negatives with online as well as FTF, and this chapter addresses both. At the same time, it is important to stay focused on the things that are difficult when going from the online method to FTF. That's the communication gap that is the focus of this book. We start with a comparison of FTF and online communication so we can understand how and when each is the correct method. In online communication the message is created and delivered so quickly that there may not be time, or thought given to barriers to communication. Online communications are characterized by fast conceptualization and forming of the message. In contrast, face-to-face interaction requires a sender getting an idea, forming it for delivery and then fully focusing on the receiver to allow message decoding. Face-to-face interactions are two-way communications, with important considerations in terms of attention to immediacy and receiver attitudes. The sender must pay special attention to physiological barriers, language, and structural design, and the risk of information overload should be low if the sender pays attention to the verbal and visual cues from the receiver. Next, the chapter focuses on ways to negotiate what is a "gap" in ability for primarily online communicators to handle a face-to-face approach. This is important, because

DOI: 10.4018/978-1-5225-1897-6.ch009

in online communications things like verbal and visual cues, emotions and environment are not major parts of the interaction.

A study published in 2011 examined a national sample of people who reported participating in face-to-face and/or online discussions, reporting that the online only group, relative to the other three groups, was less knowledgeable, less trusting, less capable of producing a desired result, less tolerant, and less interested in politics. This group also had lower trust than those who did not deliberate at all. An education-based study is covered in the chapter, discussing how students from two institutions stated reasons for selecting the online format for courses instead of face-to-face, with many noting that there were lower levels of instructor presence in online courses. Students cited two reasons for their online preference: they appreciated the flexibility and convenience and some felt that they could use their learning time more efficiently (Jaggars 2014). Many students said they learned better in the face-to-face courses. Many students simply felt they didn't learn the material as well when taking the course online, emphasizing their need or want for student-instructor and student-student interaction. The chapter demonstrates that communicators must realize that loneliness is present and work to address it. We cover hyper-personal interaction, social presence limitations, and lack of full-range communications in online interactions that may still allow for intense relationships, reciprocation, and strong involvement. As we completed a limited literature review, it became clear that several of the studies had relevant findings in terms of non-verbal communication. The bottom line is that non-verbal communications convey information about a lot of things we need to know. The chapter briefly covers crisis communication activities, stressing the need to be prepared for FTF and online communications. Finally, we ask the question when is the right time to forego online and use FTF, or oral, communication? One example is when the receiver is not really interested in getting the message. This is when FTF senders can employ verbal and visual cues to seek and nurture opportunities to get and keep the receiver's attention, allowing message delivery.

Additional Communication Considerations

So which is the right way to go, face-to-face or online? Both are needed and useful, but this book is about being able to move from online to face-to-face without undue difficulty. Face-to-face is complicated, sometimes unfamiliar, and tends to have many moving parts. Online tends to be easy for various reasons.

Let's talk about the latter and then the former. In online communication the message is created and delivered so quickly that there may not be time or thought given to barriers to communication. The sender probably doesn't even consider verbal and visual cues in delivering the message, and emotions and the environment may or may not come into play. Very often, the next message or messages come before the receiver has encoded the first one. But the digital conversation doesn't pause for that action to happen.

Online communications are characterized by fast conceptualization and forming of the message. Whether feedback is slow, fast, or nonexistent does not necessarily hinder the communication activities of the sender. The digital conversation requires great attention to immediacy. Attitudes of the receiver may or may not be considered in sending messages. The sender does not have to focus on things like physiological barriers and structural design, but must at least acknowledge the influence of language. Information overload is a big risk, especially if there is no feedback or supporting information to determine whether the audience has had too much information flow.

In contrast, face-to-face interaction requires a sender getting an idea, forming it for delivery and then fully focusing on the receiver to allow message decoding. We take verbal and visual cues, emotions and the environment into consideration and, when done correctly, we adjust the communication on the fly. We seek feedback and we work to address barriers to communication in real time.

Face-to-face interactions are two-way communications, with important considerations in terms of attention to immediacy and receiver attitudes. The sender must pay special attention to physiological barriers, language, and structural design, and the risk of information overload should be low if the sender pays attention to the verbal and visual cues from the receiver.

The matrices in Chapter 7 compare social networking and interpersonal communication characteristics between face-to-face and online interactions. They show the focus on the receiver and the signals that are available to help the communication in terms of flexibility and adjusting the meaning of the message when necessary. For online communication, the table demonstrates that the interaction is characterized by fast-paced creation and delivery with short bursts of activity and no firm requirement to seek or review feedback.

Face-to-face interactions rely on the success factors of comfort, control, confidence, timing, and focus mentioned earlier. Online connections take advantage of social capital and EVR – expectation, value, and return on investment of time. Now, we focus on ways to negotiate what is a "gap" in ability for primarily online communicators to handle a face-to-face approach. This is important because in online communications things like verbal and visual cues, emotions and environment are not major parts of the interaction. Further, online communicators can go for long periods without having to interact face-to-face.

A 2003 study examined 64 undergraduate students in 32 pairs and assigned them to a conversation with a partner they had never met (Mallen, Day et al. 2003). The conversation was either via an Internet chat program or in a face-to-face setting. The study examined emotional understanding, self-disclosure, closeness, and depth of processing.

The study found that overall there were more benefits for face-to-face communicators than achievers the case for online. The face-to-face participants did more self-disclosure in the communication, felt more satisfied with the interaction, and achieved a higher level of closeness or interconnectedness with their receiver than the online group. Interestingly, the study found that there was no real difference in judging or discovering the communication partner's emotional state when considering the two methods. However, the study results show that the relationship between the communicators developed more in the face-to-face group.

The online groups reported more conflict than existed in F2F (what this book calls FTF) interaction, an outcome that the researchers attribute to the lack of contextual cues in online communication. In apparent agreement with many experts in this field, the findings suggest that the lack of verbal and non-verbal cues causes conflict between individuals (Kraut, Lewis et al. 1982; Coleman, Paternite et al. 1999).

It could be that participants in the online condition were having difficulty deciphering the message behind the words on the screen without such cues as their partner's tone of voice. Another possibility is that participants in the online group felt freer to disagree with one another because they did not have to face their partner directly. It would seem easier to argue or contradict another person if you did not have to look them in the eye. (Mallen, Day et al. 2003)

A study published in 2011 examined a national sample of people who reported participating in face-to-face and/or online discussions to determine process, effects, and motivations (Baek, Wojcieszak et al. 2012). This study was based on primarily political discussions or deliberations, but the findings are relevant for our communication discussions. The participants were in one of four groups for this study: people who deliberated FTF and online, people who only deliberated online, people who only deliberated face-to-face, and people who did not deliberate. The study found that the online only group, relative to the other three groups, was less knowledgeable, less trusting, less capable of producing a desired result, less tolerant, and less interested in politics. This group also had lower trust than those who did not deliberate at all.

The study findings also reported that reaching consensus was less important to online deliberators than it was to offline communicators in terms of an issue, its causes and how to address it. The inference is that face-to-face deliberation is more likely to result in consensus. Face-to-face deliberators in the study were more consensual in terms of the participant's agreement with the group decision and their perception that others in the group were in agreement. Face-to-face interactions were found to lead to actions to address the problem more often than online interactions.

An education-based study published in 2014 evaluated the course choices that students make (Jaggars). Community college students from two institutions discussed reasons for selecting the online format for courses instead of face-to-face, with many noting that there were lower levels of instructor presence in online courses. Students cited two reasons for their online preference: they appreciated the flexibility and convenience and some felt that they could use their learning time more efficiently (Jaggars 2014). Some study participants noted that they had to teach themselves the course material because there were lower levels of instructor presence. Students also seemed to prefer to take what they considered to be easy courses online, but preferred to take important or difficult courses face-to-face. Some students did prefer online courses because, for instance, they were older and wished to avoid or minimize face-to-face interaction with other students. As one older student explained:

I think a lot of the older, mature people take online classes because they are afraid of the classroom. I was when I first took my first class. I'm like, 'I'm the oldest thing in here and these kids just got out of high school. I can't remember all of this stuff.' And I think the older person, the mature person, leans toward the online classes basically because of, you know, it's almost like stage fright. I mean being out of school for twenty years and then going back to a classroom, it just kind of scares you. It did me. (Jaggars 2014)

There was a lot of support in the study for face-to-face communication. Most students stated that they learned better in the face-to-face courses. In fact, students said they took face-to-face courses to experience a strong connection with the instructor and to stay connected to the campus and their peers. The bottom line was that many students simply felt they didn't learn the material as well when taking the course online, emphasizing their need or want for student-instructor and student-student interaction. The researchers suggest that colleges start to take a hard look at providing some answers to these findings because if not, they may begin to see online registrations taper off. In this study, students seem to need more of the human touch in learning.

Several studies of loneliness in college students point out difficulties in communicating face-to-face, or even reluctance to try to communicate in that way. Those studies point out that the use of social networking site by students to communicate with their parents resulted in increases in the level of emotional loneliness, anxious attachment, and conflict in the parental relationship (Moody 2001; Kim, Larose et al. 2009; Gentzler, Oberhauser et al. 2011; Brown 2013). Anxious attachment is about intimacy and closeness as core needs of a person. Because of these needs, the person seeks reassurance that things are fine and that they have ready access to their partner. This is important because the anxiously attached person needs their partner to be emotionally and physically available to satisfy their needs, whether good or bad. Very small shifts in the relationship can have very serious effects on the anxiously attached person. This explanation connects to feelings of loneliness, and the studies examined above suggested that lonely students use the internet more, causing negative life outcomes. Lonely students tried to replace satisfaction that could not be obtained in the real world with that gained from the benefits of internet and social networking sites.

A published thesis studied the social skills and anxiety and technology use and preferences of college students (Brown 2013). The analysis discusses the ever-growing technological communication in America and the negative and positive outcomes that have resulted from that growth. There is mention that some employers complain about the lack of interpersonal communication skills they experience in interviews with prospective employees. This results in a renewed focus on communication skills as one of the top qualities employers seek (McKay 2005). The published thesis demonstrated some degradation of social skills in students who preferred technological communication over face-to-face. As measured on a social skills inventory and a conversational skills rating scale, the lower social skills belonged to students who preferred online activities. The study characterized the deficiency as lack of social control, which means exhibiting less tact and self-confidence in social situations. These students also struggled more than their counterparts with non-verbal social skills.

The results of an online survey in 2011 demonstrated that more than 200 college students had challenges in the area of online communication and loneliness (Gentzler, Oberhauser et al.). The students were asked about their interactions with the parent they identified as their closest family member. The students with the more satisfying, intimate, and supportive parental relationships reported that they had more frequent phone conversations. In contrast, students who primarily used social networking sites to connect reported higher levels of conflict in the parental relationship, as well as increased levels of loneliness and anxious attachment. Phone conversations are certainly not face-to-face interactions (unless done via Skype or Facetime, for instance), but they provide more human contact than strictly online interactions. So the online study referenced here is valuable to our continuing discussion of online versus face-to-face communication.

More than 2,000 Americans were surveyed in 2000 to assess the impact of internet activities on loneliness. The study found, in part, more loneliness evident in people who have online friends than in those who do not (Coget, Yamauchi et al. 2002). The limited literature review in this chapter indicates two things. First, loneliness happens to people who are involved in online communications. Second, it is not something that can't be addressed, mitigated, or resolved. The bottom line is that communicators must realize that loneliness is there and work to address it, especially as it relates to strengthening interactions and promoting good health and happiness in people.

The study of loneliness lets us understand that people try various methods to present themselves in a certain light, called impression management. They do this because they are looking for what's appropriate for a certain social context (to be liked, to dominate, or to induce fear). Those who use this approach should avoid any behavior that resembles social manipulation or false self-presentation. Those are unacceptable ways to achieve social advantage. In addition to impression management, people create cyber identities that differ, sometimes significantly differ, from their real identity. The fact that social cues are reduced allows the sender to control the verbal content of information and create personal attributes that do not exist. Impression management becomes simplified when the sender starts "(a) Trying to transmit socially favorable personal information, attitudes and values while omitting socially disadvantageous personal data; or (b) engaging in social deception by conveying socially agreeable but untrue or incorrect personal information." (Bubaš 2001) Impression management offers online communicators the ability to mask gender, age, and ethnic differences if they choose. Some experts believe this makes quality of the interaction, task expertise, and work related effectiveness greater assets to the overall communication. Similarly, those who do not do FTF well, or do not want to, can function better without having to communicate on an interpersonal level. This personal control by sender over what receivers "see and hear" can offer more interpersonal satisfaction and mutual attraction in group or team settings. These are benefits online communicators consciously or unconsciously seek out every day.

In 2001, Goran Bubaš discussed hyper-personal communication, which is communication that appears more desirable than what we tend to experience in analogue FTF interactions (Walther 1996; Bubaš 2001). Hyper-personal is different than impersonal communication, which features low levels of social interaction and is task oriented. Interpersonal communication is, as we've learned through our journey here, defining ourselves through social interaction to create and negotiate identities, relationships, and meanings. Hyper-personal communication is characterized by idealized perception of message producer(s), optimized self-presentation, channels that support information management, and feedback loops that allow the activity to increase in intensity with minimal cues (Bubaš 2001).

Through hyper-personal interaction, social presence limitations and lack of full-range communications in online interactions may still allow for intense relationships, reciprocation, and strong involvement. However, there may be sacrifices in the communication atmosphere that come from impression management. Communicators may be disillusioned or disappointed when their online interactions don't translate well to face-to-face situations. This is because the relationship development and cohesiveness created online may be idealized, and they may find errors or fabrications in the physical and cultural characteristics of the actual communication partners. These issues could decrease team performance, cohesion, and collaboration when there are discrepancies in the original understanding of identity.

Impression management and hyper-personal communication are powerful, effective resources that people use every day to communicate online. There are many very positive outcomes as stated before. However, these two constructs are related to some of the downsides of online communication that we see and talk about every day. These realities may lead to what psychologists call behavioral disinhibition. Adam Joinson defines it as "behavior that is less inhibited than comparative behavior in real life and that is characterized by an apparent reduction in concern about socially favorable self-presentation and the positive judgement of others." (1998) Behavioral disinhibition happens when people have a lack of restraint and it becomes reality in the form of poor risk assessment, impulsive actions, and disdain for social conventions. The outcome can be like mania in some ways, with aggressive outbursts and disregard for common courtesies.

When communicating FTF, we follow some rules in terms of immediate negative feedback, interpersonal interaction, and visible inappropriate behavior to be socially appropriate and avoid possible sanctions socially. But in the online environment, there may be a false sense of security based on the anonymous, physical safety of the room where we communicate. We are far removed from those who might filter our inappropriate communication and we are hidden in the identity and personality we've chosen for ourselves. Online communicators don't face the same pressure as in FTF communicators in terms of facing negative consequences of potentially inappropriate and damaging behavior. In this way, it is easier to exhibit anger, aggression, self-disclosure in poor taste, or unsavory material like pornography. The freedom that online communicators feel through disinhibition, impression management, or hyper-personal communication can lead to issues. The issues covered here are flaming, aggression, inappropriate self-disclosure, risky computer mediated communication, and problems in constructed reality.

Disinhibition-Related Issues in Online Communication

One issue that causes problems is called flaming. Flaming is an online interaction that turns hostile with insulting messages, or flames, between sender and receiver. We see this happen in internet forums, UseNet groups, game lobbies, social networks, and chat rooms. What happens is that people from all walks of life, with differing ideologies and different cultures, who are also different ages and from different generations, come together with no real bond. Because there are no filters in terms of verbal and visual cues, people may be uninhibited in saying what they think or feel. They may not know or care how it will affect the receiver, and the interaction turns volatile.

Such episodes of verbally inappropriate behaviors are relatively common in discussion forums on the internet, but can also occur in e-mail messages and subsequently disrupt relationships or team atmosphere. Both the lack of 'social presence' and immediate FTF feedback, as well as the potential of CMC for communicating an impulsive message when the receiver is unavailable for other means of contact, can lead to flaming in work groups or teams. (Bubaš 2001)

The next issue we find is aggression that arises when people are frustrated by an outcome they didn't expect, they don't want, they don't value, or that they want to change. They can be upset with the technology or the mechanism by which they connect, or their anger and frustration can be the result of what they deem disrespect on the part of the communication partner. They might feel there is an unnecessary amount of time being spent waiting to get a response. This is, of course, unreasonable in some cases because the message may have not gotten through, the receiver might be absent or ill, the message may be unclear, or the receiver may choose to ignore it altogether. In *Psychology of the Internet*, Patricia Wallace explains that frustration is more likely to bring out an aggressive response when we are close to achieving an important goal and something, or someone, blocks us from achieving it (Wallace 2015). Aggression can take over the emotions of the sender and the message or response is released to the airways with one click, almost without thinking. If the aggression causes aggression in the receiver of that message, the problem is made worse. We are faced now with road rage on the internet. These are hard emotions to control in face-to-face interactions, but online they seem to take on a life of their own. Why? Sometimes online communicators feel like invisible combatants whose words hurt and whose armor is indestructible. We must understand that the feelings created by flaming and aggression tend to be real and they exist, even if we can't experience them for ourselves.

Inappropriate self-disclosure is really more about the damage that can be done to the sender than about damage that the sender can visit on others. This is about being careful in what you post about yourself. A short bio or curriculum vitae is normally sufficient, so be careful about posting more. Keep in mind that e-mails and posts can leave digital signatures, so senders must realize that they can be found by those who care to look. Once posted, your information stays forever and it can be very hard to correct the record. You can spend days reading articles on the internet about people who are victims of wrong or inappropriate information and who can't get it corrected or removed. Sometimes the person thinks they've wiped it from all memory, but someone finds the link and posts it to YouTube as an "epic fail." The bottom line is to be a good steward of your own information. Make sure the self-image you portray is actual or at least defensible. The first place employers, lawmakers, and solicitors check is your online profile. Make sure it is a kind and accurate depiction of you.

Risky communication technology is about privacy and protection. Online communicators are sometimes out of touch with the reality that there are criminals and peeping Toms just waiting to execute a privacy violation. Don't help them! Too many times internet providers allow the private information of their users to be compromised, but online communicators also reveal a lot about themselves that they don't need to disclose. Users should take care about the sites they visit and guard their personal information. Unfortunately, these dangers are largely insufficient inhibitors of risky behaviors on the internet. But this is a good place to cover them and provide some perspective.

Constructed reality problems are important to understand. Constructed reality, discussed in Figure 1, is really social constructionism; a theory in sociology and communication circles that examines the development of jointly constructed views of the world. These views become the basis for shared assumptions about reality. In simpler terms, constructed reality is a person's view of the world that is created by perception, supposition, and desire. There may or may not be a basis, but this the foundation by which people communicate. So people build the world in which they operate according to this theory.

Every day, we see or hear about an internet hoax, with someone pretending to be someone they're not, or a story of incredible kindness where a stranger gives someone a $200 tip. You've heard many of these. This is why we have and abundance of fact-checking and myth-debunking sites like the

Figure 1. A note on social construction of reality
(SparkNotes 2006)

A Note on Social Construction of Reality

Our reality is also a complicated negotiation. What is real depends on what is socially acceptable. Most social interactions involve some acceptance of what's going on. While we participate in the construction of reality, it's not entirely a product of our own doing.

> *Example: A wealthy individual, whose basic survival needs are met many times over, buys his pets gourmet, organic food that costs more per week than the weekly earnings of a minimum-wage worker. He is proud that he is able to take such good care of his animals and insists that it's the right thing to do if one really loves one's pets. After all, his vet was the one who recommended that he buy that brand. A minimum-wage worker who loads that food into the rich person's car might feel anger when he realizes how much money this individual spends on his pets. The minimum-wage worker might fume that this man's pets eat better than he does. He might wonder whether this rich man has any concept of reality.*

four listed below. But first, a note of caution; these sites are provided for information only and there is no acknowledgement of their value, productivity, or the accuracy of what the site can do herein, only their existence. You can find more online.

- FactCheck.org is a nonpartisan, nonprofit "consumer advocate" for voters that aims to reduce the level of deception and confusion in U.S. politics.
- Snopes.com is an online resource that debunks urban legends and rumors on a huge number of topics.
- TruthOrFiction.com is an email reality check that offers to provide the truth about rumors, inspirational stories, virus warnings, hoaxes, scams, humorous tales, pleas for help, urban legends, prayer requests, calls to action, and other forwarded emails.
- Hoax-Slayer is dedicated to debunking email hoaxes, thwarting Internet scammers, combating spam, and educating web users about email and Internet security issues.

Non-Verbal Communication

As we look at Figure 2, we should understand that several of the studies we've covered had relevant findings in terms of non-verbal communication, so this is definitely an area worthy of more analysis. There are a number of things

Figure 2.
© Can Stock Photo Inc./radiantskies

we have in mind when we refer to non-verbal communication. We can start with body movements, which are sometimes call kinesics, like nodding or shaking the head or hand gestures. Kinesics also includes posture -- whether people cross their arms or how they stand or sit. Non-verbal skills cover eye contact because it can signal the level of trust and trustworthiness of the communication partner. Para-language refers to aspects of the voice that are not considered speech, like pitch, tone, and speed or rate of speaking. Closeness or personal space, sometimes call proxemics, determines the level of intimacy of the interaction, and facial expressions like smiling, frowning, or blinking are non-verbal. Finally, non-verbal communication includes physiological changes like blinking or sweating, indicating a person's level of nervousness. All of these non-verbals are important to good communication, because without them we lack the ability to convey some important social cues. Without them, it is problematic to accent or stress certain words or to express tone or feeling. These are things we rely on in face-to-face communication but that are largely unavailable online. That's why social media and other online communicators may be perceived as emotionless, cold, or angry when sending e-mails, instant messages or texts. Smiley face emoticons are the online way of addressing this void to stress the importance of an idea, show happiness or anger, or signal wild agreement (Grice 1970).

Non-verbal communication is so important because it can be a truer measure of the person's feelings or state of mind than the words spoken or the text sent. They tend to be involuntary or uncontrollable. Senders and receivers should pay attention, because non-verbal communication tends to be more emotional and basic, leading to instinctive views of the actual communication. When the verbal and non-verbal are a mismatch, senders should trust the non-verbal response more than the words used. When there is no non-verbal message, the person may be controlling their body language to hide their true emotions or beliefs.

Reading non-verbal communication is challenging because it does not have a fixed meaning. Non-verbal communication is driven by the context in which it occurs in terms of the people involved, the location of the interaction, and the culture. A nod between two strangers is different than a nod in a business meeting, and even more different between two people in a group who may want to quietly leave. It can be very, very different in business and social situations. It can be compelling in a crisis. The other interpersonal communication complication is that without some other activity being involved, non-verbals are hard to interpret. Gestures and expressions are not exchanged in a void; any number of other things are going on. Non-verbal communication consists of a complete package of hand and eye movements, expressions, gestures, and postures that senders and receivers are constantly interpreting in combination with speech.

Figure 3 offers additional perspective on non-verbal communication.

The bottom line is that non-verbals convey information about a lot of things we need to know when we communicate. I may nod my head yes to your question, but if I'm frowning at the same time you may wonder about my sincerity. I may be sending you a mixed symbol, possibly saying "I'll agree so we can move on, but I don't like it." Non-verbals modify or reinforce what we say and demonstrate our emotional state in a lot of ways. They make statements about relationships, defining or reinforcing the terms and status, and they provide feedback on the communicating parties. They can also signal the beginning or end of a conversation in the way they can regulate the flow of the interaction.

Figure 3. Examples of culture-specific non-verbal communication
© SkillsYouNeed (2015)http://www.skillsyouneed.com/ips/nonverbal-communication.html.

Examples of culture-specific non-verbal communication

1. The popular stereotype of Italians, involving big gestures, lots of hand-waving, and plenty of loud and excited shouting, may be a stereotype, but it exists for a reason. In the Italian culture, excitement is shown a lot more obviously than in the UK, for example, and non-verbal communication tends to be a lot more obvious. This can make it much harder for Italians to interpret non-verbal communication in the UK or USA, where it is more subtle. However, even in Italy, there are geographical variations.

2. The thumbs-up gesture, which generally signals approval in English-speaking countries, is considered offensive in other countries, including apparently Greece, Italy and some parts of the Middle East.

3. Making a circle with your thumb and forefinger like this means OK in Western cultures. It is used in particular by divers in this way. In Japan, however, it is reputedly the sign for money, and in Arabic countries, it is a threat.

It's worth being careful how you use gestures and body language!

Crisis Communication Applications

When a crisis happens, there is often no time to react. Despite our current analysis of FTF and online communication, and the communication gap that exists between them, social media will likely continue to take the lead in disseminating information. So it's beneficial for our learning journey to have a crisis communication analysis. In crisis, online communications tend to be faster than ever, but this is also a time when they may be most inaccurate. Competing stories, self-proclaimed experts, and, most importantly, concerned families and friends are searching for information. At some point, there is a need to have face-to-face conversations to disseminate information. How often have you seen a press conference go wrong in these situations? How often do officials struggle with saying the right things to the victims and their families with empathy and understanding? These are situations when reliance on online communication can be the enemy to effective interactions and information sharing.

An effective social media approach is important, but it is also important to balance the online method with good, two-way communication practices for the inevitable face-to-face interactions that will occur. There are some basics that are important. Get the story out first instead of waiting on the media to break the news. The best way to make your point is to make it first. Craft your messages in advance and tell the truth when the crisis begins. If you don't know, say so. Stick to the event and to the facts, staying in your lane to share and discuss only information that you know to be true. Manage your emotions when speaking. Correct the record when someone has the information wrong. These are all time-tested suggestions, but they become harder when the event happens and everything is occurring at warp speed.

It is helpful to have a plan to alleviate stress and keep composure when communicating in a crisis. Various people are sharing information, ideas, and opinions, trying to help when people want answers. People also want to be connected to someone or some organization that can help.

Figure 4 spells out my four pillars of social media, which can help provide the basis for providing answers that matter, both in online and FTF situations. The goal of communication in this sense is to deliver messages of shared value using two-way communications. Communication and collaboration are methods that ensure continuous communication occurs, while education leverages learning styles and venues. Inform is the method that will focus your messaging on emotion, empathy, and sympathy to facilitate communication. To clarify, this fourth category, inform, was originally cast as entertainment, but has been revised to be more appropriate for a crisis communication use and value.

Figure 4.

4 Pillars of Social Media

Understanding Their Meaning in Crisis Communication

Communication

EXPECTATION: Two-way, unfettered communication
VALUE: Receive information that satisfies a need
RETURN: Create an "open door" for information

Collaboration

EXPECTATION: Two-way, unfettered collaboration
VALUE: Receive collaboration that satisfies a need
RETURN: Create an "open door" for collaboration

Education

EXPECTATION: Educate someone or some community of interest
VALUE: Satisfy a need for information that is more difficult or time-consuming to get elsewhere
RETURN: Leverage different learning styles and venues

Inform

EXPECTATION: Make my day better, or less bad, with your message
VALUE: Ease my mind
RETURN: Create an "open door" for the public to ask questions

These tools and recommendations can be very valuable in crisis situations. Communication obviously goes beyond the challenges of online communications, attention to non-verbal communication, and good practices in crisis situations. The cues that assist communication also include, for instance, the fact that one may not be able to recognize sarcasm or humor in the interaction. These emotions can be very valuable when judging how well the person perceives the conversation, how well they understand the conversation, how willing they are to continue to engage in the conversation, or whether they will provide feedback about the conversation.

Also, the article by Bubaš gives us an excellent view of competencies that can be incorporated into computer mediated communication (CMC) to assist with and partner in FTF communication. He discussed desired online outcomes that can and should be achieved and listed them as criteria for improvement if online communicators don't have success with them (2001):

- **Efficiency:** The investment in time, people and resources in relation to how much of the planned communication results were achieved;
- **Understanding:** The level at which the message was interpreted in relation to how it was intended by the sender;
- **Appropriateness:** The degree to which CMC was fitting and acceptable in the given social and professional context; and
- **Satisfaction:** Both the sender and receiver of the message have a feeling that positive expectancies of CMC have been fulfilled.

Senders use their personal CMC skills to make sure that the medium, message, and context create outcomes based on the competencies above. This approach should allow online communicators to get some of the "human" results that are more specific to FTF interactions.

We have examined a great deal of pluses and minuses concerning online communications. But when is the right time to forego online and use FTF, or oral, communication? We start with situations where the receiver is not really interested in getting the message. This is when FTF senders can employ verbal and visual cues to seek and nurture opportunities to get and keep the receiver's attention, allowing message delivery. Next, understand the importance of feedback and take advantage of the ease with which it can be obtained when one can observe facial expressions, ask questions, and observe other non-verbal behaviors. Emotions that are out of control can cripple communication, so being face-to-face can provide the necessary indications that will allow sender and receiver to take a step back, cool down, and find a better time and place for continuing the activity. The 4Rs of New Age Communications can be very helpful in these situations. Oral communication is proper when the receiver won't read because they are uninterested, busy, or preoccupied.

FTF communication can provide increased flexibility and opportunity for emphasizing the message when the intent is to persuade or convince. Take the chance to listen as well as the opportunity to remove resistance and change attitudes. Discussion can be achieved for a complicated subject in FTF interaction, allowing understanding of complex messages. FTF communication is recommended when, for whatever reason, the message criticizes the receiver. This method allows the use of verbal and non-verbal skill and observations to communicate without arousing resentment, and this method tends to be less threatening because it isn't formalized in writing. And of course, oral communication is the way to go when the receiver prefers one-to-one contact.

In the next chapter, we will discuss how we bridge the "gap."

REFERENCES

Baek, Y. M., Wojcieszak, M., & Delli Carpini, M. X. (2012). Online versus face-to-face deliberation: Who? Why? What? With what effects? *New Media & Society*, *14*(3), 363–383. doi:10.1177/1461444811413191

Brown, C. (2013). *Are We Becoming More Socially Awkward? An Analysis of the Relationship Between Technological Communication Use and Social Skills in College Students.* Academic Press.

Bubaš, G. (2001). *Computer mediated communication theories and phenomena: Factors that influence collaboration over the Internet.* 3rd CARNet Users Conference, Zagreb, Hungary.

Coget, J.-F., Yamauchi, Y., & Suman, M. (2002). The Internet, social networks and loneliness. *It & Society*, *1*(1), 180.

Coleman, L., Paternite, C., & Sherman, R. (1999). A reexamination of deindividuation in synchronous computer-mediated communication. *Computers in Human Behavior*, *15*(1), 51–65. doi:10.1016/S0747-5632(98)00032-6

Gentzler, A. L., Oberhauser, A. M., Westerman, D., & Nadorff, D. K. (2011). College students use of electronic communication with parents: Links to loneliness, attachment, and relationship quality. *Cyberpsychology, Behavior, and Social Networking*, *14*(1-2), 71–74. doi:10.1089/cyber.2009.0409 PMID:20973676

Grice, H. P. (1970). *Logic and conversation.* Academic Press.

Jaggars, S. S. (2014). Choosing Between Online and Face-to-Face Courses: Community College Student Voices. *American Journal of Distance Education*, *28*(1), 27–38. doi:10.1080/08923647.2014.867697

Joinson, A. (1998). *Causes and implications of disinhibited behavior on the Internet. In Psychology and the Internet: Intrapersonal, interpersonal, and transpersonal implications* (pp. 43–60). San Diego, CA: Academic Press.

Kim, J., Larose, & Peng. (2009). Loneliness as the cause and the effect of problematic Internet use: the relationship between Internet use and psychological well-being. *Cyberpsychology & Behavior: The Impact of the Internet, Multimedia and Virtual Reality on Behavior and Society, 12*(4), 451.

Kraut, R. E., Lewis, S. H., & Swezey, L. W. (1982). Listener responsiveness and the coordination of conversation. *Journal of Personality and Social Psychology*, *43*(4), 718–731. doi:10.1037/0022-3514.43.4.718

Mallen, M. J., Day, S. X., & Green, M. A. (2003). Online versus face-to-face conversation: An examination of relational and discourse variables. *Psychotherapy (Chicago, Ill.)*, *40*(1-2), 155–163. doi:10.1037/0033-3204.40.1-2.155

McKay, J. (2005). *Employers complain about communication skills. Pittsburg Postgazette.*

Moody, E. J. (2001). Internet Use and Its Relationship to Loneliness. *Cyberpsychology & Behavior*, *4*(3), 393–401. doi:10.1089/109493101300210303 PMID:11710265

SparkNotes. (2006). *SparkNote on Identity and Reality*. SparkNotes.com.

Wallace, P. (2015). *The psychology of the Internet*. Cambridge University Press.

Walther, J. B. (1996). Computer-mediated communication impersonal, interpersonal, and hyperpersonal interaction. *Communication Research*, *23*(1), 3–43. doi:10.1177/009365096023001001

Chapter 10
Key Strategies for Communication

OVERVIEW

This chapter presents key strategies for communicating. Communication should focus on trust to build relationships that allow messages to flow from sender to receiver, and that promote feedback. We revisit EVR and social network engagement to get a clearer focus for our continuing interactions. EVR is important because it addresses ways to improve communication and focuses us on reaching multiple receivers and managing the requirements of group dynamics. When people believe that they will receive benefits from an interaction they are inclined to join the communication. We examine trust, uncertainty, and commitment. Where there is trust, there are risks. Organizational risks may include reputational damage, compliance with regulatory requirements, information leakage, loss of intellectual property, malware attacks, copyright infringement, and privacy that may be harmful. Trust is necessary to the success of any relationship. Trust is built between communicating parties when emotions are shared, and trusting relationships help eliminate uncertainty through effective two-way communication. Shared value, common understanding, and collaborative discussions are keys to building trust. The sender or receiver can be uncertain about the identity of the other party or about the behavior of the other party. Behavioral concerns can include intended behavior like cheating or withholding information or unintended behavior like missing the communication session due to illness or being caught in traffic. Everyone experiences some level of uncertainty in communicating. Commitment is giving time and attention to the communication and taking care with the

DOI: 10.4018/978-1-5225-1897-6.ch010

message, telling the truth, seeking shared value, and staying engaged to seek feedback. Communicators who are committed send the right message with the right delivery for effective interactions. Committed communicators will do everything in their power to send the right message with the right delivery. Commitment is valuable for effective interactions. Across generations, we must regain the ability to look each other in the eye and communicate. We have become increasingly desensitized to the human element. The FTF communicator has the power to influence the conversation and motivate a person or a group to achieve a goal or goals. Moving from online to FTF communication requires awareness to connect and then analyze a person or audience and sharing information that matters. This method of communication is basic, relevant, high-touch, powerful, complementary, interactive, timely, compelling, and committed. The chapter will give explanations for each of these claims to facilitate effective communication.

Building Trust in Communications

A developmental focus was introduced in the late 1980s (Zucker 1986), establishing three central mechanisms of trust production: process-based, characteristic-based and institutional-based. Process-based focuses on reciprocal, recurring exchange, characteristic-based is defined by social similarity, and institutional-based is determined by expectations embedded in societal norms and structures.

Building trust is rooted in perception in many ways. People who believe that they will receive benefits from the interaction are then inclined to join the communication. The more they believe their current behavior contributes to some end state they value, the more active they will be in the communication. The benefit can be real or perceived and it doesn't matter which one the receiver values as long as they value it. The benefit could be valuable information, the availability of new challenges, or the chance to make new and valuable connections. Along with trust, there is some risk. Risk is affected by barriers to communication. In the previous chapter, the discussions of disinhibition, impression management, and hyper-personal communication suggested risks that reside in many online communications. But there are other risks in communicating online or otherwise. The organizational risks may include reputational damage, compliance with regulatory requirements, information leakage, loss of intellectual property, malware attacks, copyright infringement, and privacy that may be harmful. Many of these risks are applicable to group communications. Damaged reputation and loss of intellectual property are also personal risks, with the former possibly affecting personal brand

and the latter possibly affecting information leakage if posting information reveals more of one's personal life than one intended. Individuals can cause their own problems with privacy, but the more dangerous outcome is when hackers cause problems or when postings are inaccurate or inflammatory.

Risk can be managed by owning the truth about reputation and through relationship building activities. Control any site you own and use strong privacy and security settings while avoiding suspicious third-party applications. Treat everything you post as public when conducting social networking activities.

Most importantly, carefully assess the people you share with, and then assess them again. In your social networking activities, you might use a "friend of a friend" system for vetting people. If they know or are connected to someone in your network that you trust and respect, then maybe you can accept connecting with them. Without that, it might not be wise to accept someone you haven't met. Having said all of that, you might want to take some risks. There are just too many people involved in the social networking adventure who are too compelling to ignore, delete, refuse, or turn away from. Once the adventure has captured your full attention, push forward with privacy controls (Brown & Schario 2014).

When communicating, craft the message carefully and choose the right delivery, but don't avoid communicating for fear that you will lose control of the message. That's why it's important to conduct full-range communications where feedback is included and there are multiple opportunities to change or enhance the communication. Some communicators worry about control of the message, but don't let it overwhelm you. Two reasons for this are:

1. You may not want to control the message because at any given time you won't have all the answers; and
2. No one ever controls the message for more than a few fleeting moments.

Stay engaged, be flexible, and do what's necessary to conduct effective messaging activities.

Trust, Uncertainty, and Commitment

Trust is a necessary component to the success of any relationship as understood with the Stephen Covey quote in Figure 1. In today's world, business relations may have to be decided almost "overnight" and many are of a temporary nature. Trust is built between communicating parties when emotions are shared to address real issues. Trusting relationships help eliminate

Figure 1.

> "Trust is the glue of life. It's the most essential ingredient in effective communication. It's the foundational principle that holds all relationships."
>
> Stephen Covey
> American educator, author, and keynote speaker.

uncertainty through effective two-way communication. Shared value, common understanding, and collaborative discussions are keys to building trust.

All relationships have trust as their foundation. If the level of trust is good, the foundation is solid. If the level of trust is low or diminishing, the foundation may have cracks. Trust is about two people, or more, being vulnerable to each other, being willing to admit mistakes, and being willing to ask for help or acknowledge weaknesses. Building trust takes time, so approach one person and one relationship at a time. Leaders of groups can increase their trustworthiness by giving trust to others. Trust is a contract, and it requires that people take risks after they sign it.

As you can see from the full list of considerations in Table 1, trust is a complicated pursuit of relationships that can endure. Trust is about bonding with other people or groups, and it often requires putting one's faith or a pending decision in the hands of someone you have no reason to have confidence in. Trust is about the expectation that another will act in an honorable and mutually beneficial manner, allowing open and honest communication. Trust literature gives us a number of factors that contribute to building trust, so we will cover some of them here.

Lynne Zucker, an American professor and researcher took a closer look in developing trust. She identified three central modes of trust production: institutional-based, characteristic-based, and process-based trust (1986). Formal societal structures that depend on individual or organization-specific attributes contribute to institutional-based trust. Characteristic-based trust is about a person's ethnicity, culture and background. Personal trust, also called process-based trust, is created through face-to-face contact. Personal trust is important because it is where relationships begin.

Table 1. Keys to building trust

Building Trust
Receiver Adjustment: Building trust takes time, so senders must be patient as receivers adjust to messages and the communication process. Recognition is one way to help the receiver accept the interaction. Careful conversations to help with adjustments are important.
Sender/Leader Power: Collaborate and make the decision or course of action more suggestion than direct order. Provide choices and explain the rationale for any decisions made.
Security: Work to manage risk and raise comfort levels. Take the time to help people understand and adjust to the activity.
Risk Tolerance: Acknowledge and confront risk. Explain what's at stake, analyze options, and provide a safety net. If things go wrong, don't place blame. Work through the difficulty together.
Match Interests: Trust comes from shared value, shared interests, and share responsibility. Take everyone's point of view into consideration and accommodate when possible. The overall goal should drive common actions and shape the culture of the relationship and/or group to the best outcome.
Personality Match: Find common ground in terms of values, beliefs, and even cultural background. Emphasis "we" more than "I."
Care about Others: Make sure your actions in communicating are good for all involved. You might have to sacrifice your own wants and desires at time to allow other interests to prevail for the good of the communication. Be fair in your actions and make sure the process is fair. Try to focus on a win-win outcome wherever possible, and be prepared to compromise when that's not possible.
Capability: Use your skill and competence to manage the communication or complete the task. Senders sometimes have to relinquish control to address issues that threaten progress.
Integrity and Predictability: Use the old adage "underpromise and overdeliver." Be honest in your successes and failures and share information about the values that drive the communication effort and sender actions.
Communicate-Communicate-Communicate: Communication should be timely and candid. The best relationships are formed without constraints on thinking outside the box.

Communication is knowledge sharing, and 2014 study is important to understanding interpersonal trust and uncertainty (Hsu and Chang). The study examined the two constructs and the results demonstrate how much knowledge sharing is enhanced by interpersonal trust and how uncertainty hinders personal interest in knowledge sharing. The study reveals that uncertainty has more effect on knowledge sharing than it does on interpersonal trust. In other words, a person might be hesitant to share knowledge, but that position doesn't necessarily hinder the person's interest in building trust. This is not surprising, because our studies tell us that trust is developed as a way to reduce uncertainty perception. Mediation between trust and knowledge sharing is influenced by uncertainty.

We covered trust definitions and theory earlier in the book, so we just need to better define uncertainty. Uncertainty can take many forms. For instance, sender or receiver can be uncertain about the identity of the other party or about the behavior of the other party. Behavioral concerns can include intended behavior like cheating or withholding information or unintended behavior like missing the communication session due to illness or being caught in traffic. A cooperating partner may be guilty of misbehavior due to distractions or "noise," causing uncertainty in observation. One can be uncertain because of the interpretation, applicability, or context of an experience. Despite these and other uncertainties, taking them into account can allow meaningful trust to grow.

Everyone experiences some level of uncertainty in communicating. At the interpersonal level, people must deal with relationships (work and personal), group issues, and work requirements. Communicators manage uncertainty via collecting information using a variety of strategies, and from internal and external sources (Kramer 2015). Strategies can be direct requests or observation. In a group setting, internal sources are peers and supervisors and external sources are family and friends. Group leaders may struggle with uncertainty that they experience when communicating with stakeholders and when dealing with environmental influences. Receivers can use former interactions to assess the degree of uncertainty that is associated with the current communication activity. This is how the receiver determines the degree of confidence they can have that the sender will complete a successful message delivery. There is also something called decision trust that signifies the level of uncertainty, and the risk associated, with a decision to communicate.

Senders can build trust and address uncertainty through commitment. Commitment is about giving your time and attention to the communication and to the receiver. It is about taking care with the message, telling the truth, seeking shared value, and staying engaged to seek feedback. This is not easy, but it can be very rewarding. Senders will find that receivers are very responsive to someone who is in tune with them. This requires someone who tries to understand their needs and to address them and who is accessible and accountable. Commitment allows the sender to collaborate and get involved in a meaningful way. That is something that people respond to and want to be a part of. Focusing on commitment leads us to some additional expert testimony.

A group of researchers call communication, commitment, and trust a triad. The group studied 244 employees from a medium-size food processing organization operating in New South Wales (Australia), exploring triad relationships (Tipu & Ryan 2011). The study measured trust on a six-item composite scale that analyzed overall beliefs. The focus was good intentions of participants and an examination of the degree of faith and/or trust in various actors in the organization. Actors included co-workers and managers at various levels of the hierarchy. The study featured a new attempt to examine relationships and draw inferences relevant to communication, trust, and commitment. The study provided a basis for the examination in the form of two studies about the triad that produced inconclusive results (according to the new research). The first study analyzed trust effects on organizational commitment, resulting in a finding that trust people had in their superiors affected their commitment to the organization (Roberts & O'Reilly 1974). The study revealed statistically significant inter-correlation between trust in superior and the satisfaction with the communication. However, the relationship between organizational commitment and satisfaction with communication was not statistically significant. The second study explored the managerial communication, procedural justice of layoffs, and employees' attitudes during divestiture (Gopinath 2000). The 244 participants demonstrated higher levels of organizational commitment due to divestiture based on the managerial communication. Communication was not found to be a predictor of employee trust. These two studies provide the foundation for the "triad" study mentioned above.

Another important study by Jan De Ridder attempted to determine whether a positive attitude towards strategic direction results from effective organizational communication (2004). De Ridder studied 791 respondents distributed across 19 work units in five organizations. The study concluded that organizations have two ways to get member support. The first is creating a sense of commitment in members, and the second is to achieve member trust in management. The study also found positive relationships based on trust and commitment that result in good internal communication. After examining the three studies, the "triad" authors determined this:

Previous research on the relationships between communication, trust and commitment has yielded mixed results. Although exploratory, the results of this study clearly indicate that while a close link exists between all of the three variables, the relationship between communication and trust is by far the strongest, followed by that between trust and commitment. These findings lend support to the strong logical assumption that trust is at the center of this triadic relationship, while commitment is the end product of such relationship. (Tipu & Ryan 2011)

All of this new knowledge tells us something we should have already known; trust is central to creating commitment and loyalty in communication. Effective communication requires it. Good communication is authentic and effective communicators work to avoid misperceptions and mistrust. Effective communication can not only create trust, it can nurture trusting relationships and allow senders to adjust future communications for continued improvement over time. Trust can also help when the communication goes wrong due to misrepresentation, misunderstanding, suspicion, dishonesty, or mind games. Senders can use trusting communication to fight anger and fear and to try to keep participants from disguising, distorting, or ignoring facts, feelings, or conclusions. The bottom line is that a positive "trust climate" leads directly to effective communication.

Committed communicators will do everything in their power to send the right message with the right delivery. Commitment is valuable for effective interactions.

Social Networking WIIFM: What's in it for Me?

We are discussing communication, but let's not forget about social networking. If we ask about the WIIFM, we will start to uncover communication expectations and values that are important to all parties. We've already covered how inspirational appeals and shared value can help senders promote trust. We've covered the implied benefit of that kind of exchange, where receivers reciprocate and increase or renew their commitment to the communication task at hand through active participation. The issue goes much deeper to the responsibility of the sender to ensure that the message of choice meets the shared expectation of the parties involved for a certain outcome of the activity, and that expected outcome must be of value to all parties. To have a good basis for this discussion, see the Figure 2 box on the next page for information on EVR.

Expectation, Value and Return (EVR)

EVR develops a winning social media plan using three keys to guide activities that grow and nurture online networks. There are some negative aspects to social media activities in terms of privacy and lack of human contact, for instance, but there are far too many benefits available to advocate ignoring the digital conversation.

Done well, people can stay in touch with family, friends and colleagues. They can build meaningful relationships with people and groups by sharing interests, community activity, and professional knowledge and experience. More than that, social media can allow friends or potential employers to locate you or offer you a new job.

There are benefits, but there is also a lot of "noise" in the channel. Distractions and barriers to communication can make it difficult to find relevant

Figure 2.

Expectation, Value and Return (EVR)

EVR develops a winning social media plan using three keys to guide activities that grow and nurture online networks. There are some negative aspects to social media activities in terms of privacy and lack of human contact, for instance, but there are far too many benefits available to advocate ignoring the digital conversation.

Done well, people can stay in touch with family, friends and colleagues. They can build meaningful relationships with people and groups by sharing interests, community activity, and professional knowledge and experience. More than that, social media can allow friends or potential employers to locate you or offer you a new job.

© Can Stock Photo Inc./everythingpossible

There are benefits, but there is also a lot of "noise" in the channel. Distractions and barriers to communication can make it difficult to find relevant information and relationships in the online space. But the kind of tools and resources offered throughout this book can help in bringing the benefits of seeing and hearing current news and conversations. There's a wealth of information and connections available, too, and senders and receivers can work together to filter out or ignore the things are irrelevant or annoying.

So now we can discuss EVR, which is the key to social networking. This is how you demonstrate the WIIFM to your intended audience, whether they are one person or a group. In the true sense of social media, and communication, interaction, people are motivated to contribute valuable information to a person or a group because of an expectation of value in return. We're talking about three things. First is the ability to get useful help in a two-way arrangement, and second is the opportunity for a valued information exchange. The third benefit to EVR is getting and giving recognition that matters to others. Of course there's more to social media participation than just understanding EVR. But EVR gives you the tools to effectively connect with others (Brown and Schario 2014).

information and relationships in the online space. But the kind of tools and resources offered throughout this book can help in bringing the benefits of seeing and hearing current news and conversations. There's a wealth of information and connections available, too, and senders and receivers can work together to filter out or ignore the things are irrelevant or annoying.

So now we can discuss EVR, which is the key to social networking. This is how you demonstrate the WIIFM to your intended audience, whether they are one person or a group. In the true sense of social media, and communication, interaction, people are motivated to contribute valuable information to a person or a group because of an expectation of value in return. We're talking about three things. First is the ability to get useful help in a two-way arrangement, and second is the opportunity for a valued information exchange. The third benefit to EVR is getting and giving recognition that matters to others. Of course there's more to social media participation than just understanding EVR. But EVR gives you the tools to effectively connect with others (Brown and Schario 2014).

EVR is important, because it addresses ways to improve communication. It also helps us understand that social networking approaches can also be helpful in FTF activities. In other words, as we continue to work on bridging the communication gap between FTF and online communication, we want to make sure we're helping people drive across the bridge in both directions. So, using a social networking approach, senders should be thinking about perceived improvement potential (PIP), which involves a conscious determination of whether the activity or commitment involved will bring personal benefits (Brown 2011). The success of social networking as it relates to performance is based on normative practices as well as on relationships that will build trust, foster interaction, and address participant needs. PIP, based on a self-perception of effectiveness, has a double purpose of predicting the intention to behave in a certain way and determining whether participants will actually exhibit that behavior. Integration of network activities, or interpersonal communication activities, work to identify social relationships in a community and its formal institutions, and seeks to ensure collaboration and cooperation. Next, understanding the nature of social relations allows senders to work on bonding and bridging activities to support the most open and straightforward messaging possible. Finally, by fostering widespread cooperation, trust, and efficiency, communicators can develop social capital that keeps partners from feeling isolated, ignored, or violated. Communicators must find value and cooperation in social networking activities to satisfy the goal and objectives of the interaction event or events. This transforms ideas into realities, promotes information sharing, and helps with network building.

A word of caution. PIP is only relevant if communication participants believe their actions will result in some level of improvement or bring some reward, regardless of how small that reward might be. PIP is about addressing resistance to change and breaking down barriers, and these actions will resolve communication challenges. Four groups of barriers to change, or blocks, were introduced by Dr. Tudor Rickards, an English consultant in creative problem solving: strategic, value, perceptual, and self-image (Marsh 1990). Strategic blocks deal with the way we implement ideas and are influenced by previous experience and habit. Value blocks deal with our approach to ideas, which are influenced by our prejudices, traditions, and upbringing. Perceptual blocks deal with task orientation and are influenced by our single-mindedness or what we are doing at the time ideas are presented, especially if we are pressed for time or under stress. Self-image blocks deal with whether people believe they have the ability to implement new ideas and are influenced by experience and the reactions of others. Understanding these barriers, and adjusting to them is important to, as Marsh (1990) says, "meeting the challenge of change."

For purposes of the book, the challenge is more about changing the way we communicate to work on the communication gap. Success relies on normative practices as well as relationships that build trust, foster interaction, and address participant needs. There are other commonalities, such as how social networking stresses focused interaction between individuals (Sabatini 2009) and how individual performance concerns task-specific interaction (Ivancevich 2008). Bonding, bridging, and linking activities contribute to a shared relationship between communication partners that can be mutually beneficial. Bonding is a relationship with family and friends and it is the strongest form of social capital. Bridging is about the relationships with friends of friends. Linking capital is the weakest form, as it is between a person and a government official, elected leader, or some other public figure.

Thoughts on Face-to-Face Interactions

The previous coverage of social capital types and the knowledge about trust and uncertainty leads us to a deeper dive into FTF communication. Across generations, we must regain the ability to look each other in the eye and communicate. We have become increasingly desensitized to the human element. I have noticed that murders and other crimes against humanity are becoming less and less shocking to hear, because we spend so much time in a social media world. Digital communication and digital events don't have emotions, human contact, or complete realism. We can basically be anybody and anything we want. However, there is still a need for human contact. For instance, most job interviews are still human to human. In this instance, if you can't do FTF, you don't get the job.

In face-to-face encounters, the sender has the power to influence the conversation and motivate a person or a group to achieve a goal or goals. We're going to talk about connected communications, which are the face-to-face discussions that create social capital and build relationships. Once again, we'll focus on active listening, group or personal collaboration, and ongoing self-assessment. We begin the discussion with some common ingredients that make a group successful:

- A goal that can be achieved;
- An agenda that all parties understand;
- Honest and honorable behavior by all players;
- Personal accountability;
- Cohesiveness; and
- Feedback and support to continue the communication.

There may be different words, but these are minimum essentials for successful communication. The sender must work on this list to create valuable information sharing. Good communication starts with an awareness of individual or group needs. Why? Everyone is interested in self-improvement. Everyone wants to be the example that others strive to emulate. Good communicators will also think about balance in their approach. Too much emphasis on self-improvement can deprive a communication partner of need fulfillment. Overemphasizing group needs can rob the individual of a sense of self and hinder the kind of creativity and innovation that every good communication needs. The key is to understand people and balance the needs of the sender with the needs of the receiver.

Senders can improve their communication by employing vision. Vision describes the desired end state for the communication. It assists in understanding necessary current and future interaction needs and courses of action. The sender's vision determines what is necessary for the parties to continue to communicate. The sender's vision guides the communication for the short and the long term, creating relationships that could result in a brighter future. Marshall Loeb and Stephen Kindel say vision comes from three places: experience, knowledge, and imagination (1999).

Experience lets you visualize from the way you live. Knowledge lets you visualize from what you've learned. Imagination springs from the randomness of life, because it synthesizes knowledge and experience, but it's also connected to desire. By taking your life experience and factoring in all the possibilities and all the ways that you can see yourself and the world, your imagination allows you to grasp the possibility that waits just over the horizon. Luck and possibilities are intertwined, because a change in luck can determine your possibilities. And possibility is the stuff of visions.

For success in FTF groups, let's look at the 12 characteristics of effective organizations as listed by Glenn M. Palmer in his book, Team Players and Teamwork: The New Competitive Business Strategy (1991). Finding these characteristics in your communication, whether group or personal, can be a good sign that you've got the potential for success. Adding those that aren't currently a part of your communication will enhance your success.

1. Clear purpose in an action plan that is defined and is now accepted by everyone.
2. Informal climate that is comfortable and relaxed with no obvious tensions or signs of boredom.

3. Participation and in-depth discussion where everyone is involved.
4. Members use effective listening techniques such as questioning, paraphrasing and summarizing to get out ideas.
5. There is civilized disagreement, and the team is comfortable with it and shows no signs of avoiding, smoothing over, or suppressing conflict.
6. Consensus decisions are used. For important decisions, seek substantial but not necessarily unanimous agreement by using open discussion of everyone's ideas, avoidance of formal voting, or easy compromises.
7. Open communication is the norm, where team members feel free to express their feelings on the task as well as on the group's operation. There are few hidden agendas and communication takes place outside of formal meetings.
8. The organization sets clear roles and work assignments for each team member. When action is taken, clear assignments are made, accepted and carried out. Work is fairly distributed among team members.
9. Shared leadership is another important characteristic. While the team has a formal leader, leadership functions shift from time to time depending upon the circumstances, the needs of the group, and the skills of the members. The formal leader models the appropriate behavior and helps establish positive norms.
10. External relations are effective because the team spends time developing key outside relationships, mobilizing resources, and building credibility with important players in other parts of the organization.
11. Diversity means the team has a broad spectrum of team-player types including members who emphasize attention to task, goal setting, focus on process, and questions about how the team is functioning.
12. Self-assessment is a periodic requirement. Stop and examine how well the team is functioning and what may be interfering with its effectiveness.

Good communication requires motivation to keep everyone interested and involved. The real challenge is to find out what makes the receiver tick. Senders should engage with receivers to determine how to deliver messages that are valuable. If the message isn't providing information that the receiver wants, the sender must make adjustments. Those adjustments may be in the form of help, advice, or guidance.

When focusing on online communication, it is customary to discuss conflict that happens in many interactions. In FTF interactions, we should focus on managing stress. Why is this important? It's important because experts say stress is always present. Communicators must adjust to stress because it can cause emotional outbursts or anxious moments or total shut down of the interaction. Think about it. Put a rubber band around a bunch of straws, with the rubber band being stress and the straws being ongoing communication. It's easy to see how important the amount of stress (stretching or tightening the rubber band around the straws) can be.

Too much stress can be a problem because it distracts both sender and receiver, causes fatigue, and can elicit feelings of mistrust. Misguided doubts can put undue stress on the communication atmosphere. Too little stress can cause laziness or apathy due to inactivity and lack of attention to detail. Very low stress levels can create a false sense of security and a laid back atmosphere, making it difficult to continue communicating.

What is stress? Stress is the body's physiological and psychological response to the demands we make on it. The demands can be environmental (noise, dust, smoke, job monotony, physical exhaustion) or psychological (a promotion, a friend's death, being fired, graduating from school). If a stress response is triggered, it doesn't matter whether the stress is caused by a positive or a negative event (Rawlins 1992). The important thing is to recognize that stress exists. Let's look a little closer at high and low stress, keeping in mind that we want balanced stress in our communications.

Many things can cause high stress. It is stressful if people don't understand the message, if they don't have time to receive it, if they are not in a comfortable place when it's delivered, or if they are angry or unhappy when they receive the news. Low stress can come from lack of communication, like if someone who just retired and is missing constant conversation or activity. Some people try to avoid stress at all costs. At any rate, it is crucial that every communicator learn how to manage stress and how to operate effectively regardless of how much or how little stress they're experiencing.

The digital conversation exists, grows, and perpetuates just fine. But there are gaps or issues when compared to FTF activities. Look at the social media examples below to illustrate the point:

Online Communication Thoughts

LinkedIn

I often get invitations from people I don't know. They are connected to me on the second- or third-level, or they found my e-mail address somewhere. I don't automatically ignore their request; I give them the benefit of the doubt. So I send them a note (if I can) asking them how we know each other. I've been at this digital conversation for more than 10 years. As of this writing, not one single person has responded to my note. So, of course, they are ignored.

Facebook

I'm a watcher on Facebook. When I discovered that I would do my PhD dissertation on social media, I signed up for everything I could think of, Facebook included. I was a watcher everywhere so I could learn each platform. Later, I became active at different levels on different platforms based on my personal and professional goals. Basically, I used my EVR beliefs to decide where and how much to participate. So I get Facebook invitations all the time for various reasons. I know most of the people who invite me. But I don't accept anyone unless one of three requirements is present:

1. They are family;
2. I want to communicate with them on Facebook, or
3. I want to see what they're saying on Facebook.

By having these rules, it helps me easily make the decision to accept or ignore. Recently, a friend sent me a text and asked why I didn't accept his Facebook invite. When I told him that I'm a watcher and that I don't really communicate via Facebook, his answer was very interesting. He said that he wanted to get me to "friend" him so I could like a business page. So I ignored his request. And that's not a bad thing in my view, because I can talk to him anytime…via text.

Virtual World

I checked out a virtual world that I learned about while I was taking public administration classes at Old Dominion University in Norfolk, Virginia. Someone invited me to join and attend a virtual meeting. She said all I had to do was create an avatar and I could come for a visit. I was going to discuss research with her, so I thought it might help secure the agreement if I participated. I called her work and personal phones to get details. She has yet to return my calls.

Facebook

People sometimes post so much, so fast, they forget what they've posted. Then, they're unprepared for the face-to-face interaction. How many times have you read a post about a trip and then asked the person how it went, only to face their surprise that you knew they travelled? How many times have you told someone you are sorry for their misfortune, only to find out they don't remember they poured their hearts out to the world?

The Way Ahead

Keep in mind that moving from online to FTF communication requires awareness. This awareness is about connecting; analyzing a person or audience, and sharing information that matters. Then, you ask questions to determine if the information was received and you seek feedback to determine the value the other person puts on the message. This information will help determine whether the communication can, and should, continue. This is a challenge, because these things don't have to be done online.

Communicating well requires engagement that makes sense of the way people connect. This approach finds ways to help communicators discover like-minded people and share valuable information with them. Senders are in search of unfettered trust where we put our lives and our communication needs in the hands of people we might not trust otherwise. We create social capital and exchange valuable content. This communication is valuable to people getting together, creating community, listening to each other, understanding the message, and providing feedback. FTF communication requires that we build meaningful relationships that will allow us to start and nurture interactions.

Engagement enhances communication by addressing the topics and challenges we face every day. Senders must ensure that they don't overwhelm the receiver causing communication meltdown that ends the conversation. The message and the environment must be managed to achieve the communication success. In addition, it's important to select network partners with care to allow you to build long-standing, mutually beneficial relationships. Shared value that creates ties that endure ensures dependable communication channels, coordinated messages, and valuable feedback.

Communication participants get enormous value from the sender's attention to these concerns and responsibilities. The value is created by the varied and unlimited points of personal and professional networks and linkages that translate to useful responses by everyone involved. Each participant can then adjust the communication appropriately. Great communicators create collaboration and/or interaction by using these valuable linkages. These concerns and responsibilities, when managed effectively, can open the door to the enormous value available to participants. The value is available as part of the process of analyzing the receiver and his or her needs for the communication activity.

Senders engage to deliver great content that advances the need to build relationships. The communication must generate interest that can be kept alive. Full-range communications are most important when we can sustain them long term and transform them into relationships.

What We Know Now

Our journey through high-touch and high-teach approaches is in its final stages. We've learned about face-to-face and online connections with people that produce effective, two-way messaging. We have cultivated three communication abilities:

1. A deeper understanding of face-to-face and online communication,
2. A focus on the gap in the communication process between the two methods,
3. And knowledge leading to resources to "bridge the gap" and create the best messaging interactions possible regardless of the method.

The first two sections, the high-touch portions of the book, covered research and knowledge about communication. The high-teach section helped us understand navigating the communication gap between FTF and online interactions. This promotes effective communication.

This work provides a new formula for education and training in communication, social media, interpersonal issues, team building, and trust for organizations and academia. The value of this work goes beyond its intended public administration focus to knowledge and resources that are applicable to other industries. The utility could be endless. The benefit of this work is that scholars and students alike have a resource that examines social networking and social media activities in a holistic process, comparing face-to-face and online interactions. This is believed to be the first substantial work dedicated to the communication gap that exists between the two methods, but that still links them together. The versatility of the work goes well beyond traditional approaches to interpersonal and group communication.

Communicators are now armed with ways to bridge the gap between face-to-face and online activities. Social networks can adversely affect communication based on the lack of human contact in day-to-day communication. We have addressed the pitfalls of the online method, including superficial communication experiences, degraded grammar and spelling, and language complications. Now, we have to characterize the substance of face-to-face communication beyond the process. We want to describe the specifics of this interaction that will help senders and receivers get the most out of their interactions.

Our educational journey seeks to return to using the full communication cycle to address the growing communication challenges posed by the evolving intergenerational use of social media. We need to reintroduce face-to-face communications into our world. Social media defines many cultural, social, and work environments of Millennials and others in powerful digital conversations. These digital conversations inherently create a cross-cultural and cross-generation communication problem in that they do not employ the full range of the communication process. Digital communication has proven itself as a communications tactic with the ability to build trust among strangers, but there are much stronger bonds that are often required for successful face-to-face interactions. As the cyber workplace grows, many managers are not able to recognize the need for individuals and staff members to have new tools to build a truly effective team. How can virtual "teams" function best when team participants are people who have never met? That first face-to-face team meeting (if it ever happens) creates a communication gap. If not successfully addressed, there is a likelihood that the team will not function as efficiently as possible. Interaction may be limited and ineffective virtual "team tools" may even doom the team to failure. There may never be a meeting of the minds that is so crucial to team and culture building.

What is FTF communication? It is a lot of things, but there are several specific characteristics listed in Figure 3 that illustrates the value of this method. FTF is *basic, relevant, high-touch, powerful, complementary, interactive, timely, compelling* and *committed (as illustrated in the puzzle inFigure 3)*. Working our way through explanations and support for each of these statements is like a puzzle, allowing FTF communicators to develop or improve skills and get the most out of their engagements:

- **FTF is Basic:** This method relies on the full communication process and emphasizes the feedback cycle, shared understanding between communication participants, and use of social cues to improve the interaction. The method focuses on feedback as a crucial element, giving the sender important clarification of the receiver's decoding and processing of the information. Mutual agreement between all parties ensures meaning and sets the stage for continued communication. Social cues give a real sense of the receiver and allow continuing creation of relationships. Digital communications are blamed for decreasing human interaction, developing superficial communication experiences, and for contributing to erosion of traditional language conventions.
- **FTF is Relevant:** The theoretical basis provided throughout our communication journey provides a foundation for communicating in a variety of situations with a variety of people. And the theoretical basis has been shown to cover FTF and online communication. Keeping the communication partner engaged is made possible with the skills de-

Figure 3.

scribed in innovation adoption, diffusion of innovations, and governing by network theories. These theories focus on behavioral changes that are beneficial to communication through positive connections based on goals, connecting with people based on identity and demographics, and creating value through important relationships. These fundamentals can assist senders to deliver value propositions that make communication meaningful. Various communication theories contribute important approaches to dealing with people. We have found that access to information is not equal, so senders can have an affect by providing messages that matter. Open and honest communication allows senders to determine the authenticity of their communication partner. This is referred to as whether or not the person is real in social presence theory, which is more relevant in online communication. In fact, social learning theory leads FTF communicators to understand continuous interaction between cognitive, behavioral, and environmental influences, thereby facilitating full-range communication.

- **FTF is High-Touch:** Effective communication is rooted in relationship building, so social capital leads us to make contact with people for information sharing. Examining connections in online communications allows us to determine strengths and weaknesses as they relate to the communication gap. Efforts to improve internal and external digital communication and to enhance organizational training demonstrates ways to positively affect productivity, levels of trust, and the ability to conduct bonding, bridging, and linking activities. High-touch is about creating and nurturing relationships through which we engage and communicate, and about ensuring feedback completes the communication cycle.

- **FTF is Powerful:** This method allows navigation of communication and keys to how we connect with people in person and online. As we compare and contrast the methods, we find strengths and weaknesses of each, but also opportunities for enhancing our communication activities. Popular communication models from Argyle, Schramm and Shannon-Weaver explain interactive processes to exchange messages and deal with noise in the channel. These interpersonal communications can define us as we realize identity, shared meaning, and relationships. Communicators express attitudes and beliefs to gain mutual understanding. This dialogue from full-range communication defines the social world as we work on self-improvement and personal relationships. The power in FTF interaction is the ability to look someone in the eye and determine whether our message is getting through. We can ask questions, we can change the approach, and we can work toward mutual agreement.
- **FTF is Complementary:** We can and should use the benefits of FTF and online communication interchangeably whenever possible. Our discussion clearly demonstrates that people can benefit from both FTF and online communication. Our communication journey focuses on the knowledge that we bridge the "full capability gap" between the two and both have value. Our high-touch approach is to take the best of both worlds and make communication better. However, we know that online communication misses, ignores, neglects, or is unaffected by verbal and non-verbal cues that give communication full substance and capability. The human element, and to how inject it into communication activities, is important for messaging and we have discovered ways to take advantage of the benefits.
- **FTF is Interactive:** The interpersonal communication theoretical basis we have created in these pages is valuable. It allows us to draw on major capabilities to conduct meaningful messaging interactions. FTF can deal with emotions, create meaning from social context, and manage personal and social identity based on the knowledge from EI, sen-

semaking, and CAT, respectively. Other theories lead in-person communicators to abilities to do the following:
  - Alleviate uncertainty;
  - Provide needed assistance throughout the communication activity;
  - Deal with interpersonal deception;
  - Adjust to communication partners who seek or avoid the issue at hand;
  - Protect privacy and control information; and
  - Create and nurture long-term relationships.

These abilities are important to communicating effectively. Senders can manage shared information through interpretation, evaluation, and decision stages of the interaction. Communication participants can collaborate to work out private information and boundaries as well as control of the shared information. Further, participants can find ways to adapt to affectionate communications or exchanges are that are emotionally charged for various reasons. They can deal with immediacy and primacy issues in feedback. They can examine the resolution of conflict to save face, correct the record, determine winners or losers, or to set terms for future interactions. Conflict, both good and bad, is part of communication and communicators must have the skills to address it.

- **FTF is Timely:** The high-teach part of this book covers a lot of ground in terms of communication capability. Communicators must assess their audience to find out what works and what doesn't when communicating. That assessment should focus on what the receiving party might consider to be valuable information. In online communications, this is often a guess, or the communication happens so fast there is little consideration of the receiver's views. When face-to-face, the communicator cannot afford to miss making an assessment or evaluation of the receiver. This assessment provides capability to arrive at a mutual agreement on the message and the sender's intent, taking into account the need to read social cues and the importance of full-range communication from sender to receiver to feedback. Two matrices are "right on time" to offer tools with which FTF and online communicators can get better at connecting, engaging, interacting, and bonding: the Social Networking Matrix and the Interpersonal Communication Matrix. Each provides the theory that is covered earlier in the book followed by an explanation of the basics of what the theory describes. Then, there are two suggested approaches each for FTF and online interactions. The introduction of the concept of Engaged Interaction offers a resource that fuels use of the triad of listen, hear, and understand, calling for senders and receivers to commit fully to the communication activity for maximum effectiveness.

- **FTF is Compelling:** We have suggested the best times and situations in which to use FTF versus online communication. Despite seeming overly critical of online communication, this book is simply advocating the best path to success. It is common knowledge that the best way to know what a person wants or thinks is to ask them, and that can't always be accomplished online. There are positives and negatives in both methods, but it is important to keep a focus on the things that are difficult when going from the online method to FTF. It's also important to take reasonable action to change a communication negative into a positive, and that's what the high-teach approach herein is all about. That's the communication gap we are addressing. Online messages are conceptualized at a fast rate and the message is subsequently formed at light speed. FTF interactions are of course slower, with the sender forming the idea, encoding it for delivery, and then paying full attention to the receiver as he or she decodes it. Emphasis on two-way interaction is the way senders seek and receive feedback. The sender must pay special attention to physiological barriers, language, and structural design, and the risk of information overload. These details are very important and require attention to verbal and visual cues from the receiver. Next, the chapter focuses on ways to negotiate what is a "gap" in ability for primarily online communicators to handle a face-to-face approach. This is important, because in online communications things like verbal and visual cues, emotions, and environment are not major parts of the interaction. Loneliness, hyper-personal interaction, social presence limitations, and lack of full-range communications in online interactions are considerations that may still allow for intense relationships, reciprocation, and strong involvement in that method. Still, the full-range communication available through FTF activities is compelling and valuable. FTF is valuable, for instance, when the receiver is not really interested in getting the message. This is when FTF senders can employ verbal and visual cues to seek and nurture opportunities to get and keep the receiver's attention, allowing message delivery.
- **FTF is Committed:** Communication emphasizes trust to build relationships that allow messages to flow from sender to receiver, and that promote feedback. When people believe that they will receive benefits from an interaction, they are more inclined to join the communica-

tion. Trust, uncertainty, and commitment are examined in communication. People build trust by sharing emotions to work on eliminating uncertainty through collaborative activities. Building trust is aided by shared value, common understanding, and deliberative conversations where there is ample give and take between the participants. As communicators work to alleviate uncertainty, behavioral concerns can include intended behavior like cheating or withholding information or unintended behavior like missing the communication session due to illness or being caught in traffic. People are committed when they give time and attention to communication interactions with emphasis on sending valuable messages, telling the truth, ensuring shared value in information, and remaining engaged to seek and respond to feedback. Every one of us can benefit from being able to look others in the eye and conduct meaningful two-way conversations. Face-to-face communicators have the power to influence important message delivery and to motivate a person or a group to achieve a goal or goals. Moving from online to FTF communication requires awareness to connect and then analyze a person or audience and sharing information that matters.

REFERENCES

Brown, M. Sr. (2011). *Social networking and individual performance: Examining predictors of participation.* Old Dominion University.

Brown, M. A. Sr, & Schario, T. A. (2014). *Social Media 4EVR: Identifying, Achieving, & Nurturing Social Capital.* Yorktown, VA: CreateSpace Independent Publishing Platform.

De Ridder, J. A. (2004). Organisational communication and supportive employees. *Human Resource Management Journal, 14*(3), 20–30. doi:10.1111/j.1748-8583.2004.tb00124.x

Gopinath, C., & Becker, T. E. (2000). Communication, procedural justice, and employee attitudes: Relationships under conditions of divestiture. *Journal of Management, 26*(1), 63–83. doi:10.1177/014920630002600107

Hsu, M.-H., & Chang, C.-M. (2014). Examining interpersonal trust as a facilitator and uncertainty as an inhibitor of intra-organisational knowledge sharing. *Information Systems Journal, 24*(2), 119–142. doi:10.1111/isj.12000

Ivancevich, J. M., Konopaske, R., & Matteson, M. T. (2008). Organizational Behavior and Management. New York: McGraw-Hill/Irwin.

Kramer, M. W. (2015). *Uncertainty and Communication in Organizations. In The International Encyclopedia of Interpersonal Communication*. John Wiley & Sons, Inc.

Loeb, M., & Kindel, S. (1999). *Leadership for Dummies*. Foster City, CA: IDG Books Worldwide, Inc.

Marsh, W. (1990). Meeting the Challenge of Change. *Australian Accountant*, *60*(1), 34.

Palmer, G. M. (1991). *Team Players and Teamwork: The New Competitive Business Strategy*. San Francisco, CA: Jossey-Bass Publishers.

Rawlins, C. (1992). *Introduction to management*. New York, NY: Harper-Collins Publishers.

Roberts, K. H., & OReilly, C. A. (1974). Measuring organizational communication. *The Journal of Applied Psychology*, *59*(3), 321–326. doi:10.1037/h0036660

Sabatini, F. (2009). Social capital as social networks: A new framework for measurement and an empirical analysis of its determinants and consequences. *Journal of Socio-Economics*, *38*(3), 429–442. doi:10.1016/j.socec.2008.06.001

Tipu, S. A., & Ryan, J. C. (2011). Communication, commitment & trust: Exploring the triad. *International Journal of Business and Management*, *6*(6), 77.

Zucker, L. G. (1986). *Production of trust: Institutional sources of economic structure. In Research in organisational behaviour* (p. 8). Greenwich, CT: JAI.

Chapter 11
New Age Communications

OVERVIEW

This book contains a communication journey that will give the reader three valuable communication abilities:

1. A deeper understanding of face-to-face and online communication,
2. A focus on the gap in the communication process between the two methods, and
3. Knowledge leading to resources to "bridge the gap" and create the best messaging interactions possible regardless of the method.

Deeper understanding of communication and focusing on the gap in the communication are high-touch issues relating to getting people to interact. Resources to bridge the gap are high-teach considerations leading to a plan for changing and improving the way we communicate. This chapter focuses on a hybrid approach to communicating that is the culmination of all we have discussed in the pages of this book.

High-Teach Approach

This chapter is high-teach, delivering a hybrid approach with resources to "bridge the gap" and create the best messaging interactions possible. The previous chapter compared the methods and timing when FTF and online communication, respectively, are appropriate for use. In many ways, this chapter

DOI: 10.4018/978-1-5225-1897-6.ch011

takes an interactionist view to demonstrate a hybrid approach that is neither exclusively FTF nor online. This is a blended approach that energizes the best of both methods, giving communicators tools and resources they have never had before. This new step in communication attempts to remove the barriers in both methods to achieve the greatest possible use of engaged interaction.

Before discussing the hybrid approach, we should remind ourselves of the resources we have already gained through the discussion in this book. Sections 1 and 2 provided a wealth of knowledge and discovery for addressing the gap in communication. The explanation of technological advances and how they caused an explosion of social media and digital communication taught us about mediated communication and how people use technology to interact with others. That knowledge has also taught us the importance of getting online communicators to see, understand, and accept the skills and benefits of face-to-face communication, and the value in understanding motivation, cultural considerations, and issues of multiethnicity. Communication theories relevant to social networking emphasize addressing the necessary improvements to identify determinants and opportunities that affect levels of acceptance and participation. Readers should now be armed with an understanding of the communication gap between FTF and online, using emotional intelligence, sensemaking, and communication accommodation theories (CAT).

Chapter 4 introduced us to a high-touch approach to social capital and to Social Networking and Individual Performance (SNIP) to pursue success in digital communication. We now know that social capital assists senders and receivers in building trust and seeking engagement. Our educational journey continued with communication methods and barriers to communication, both of which can be addressed by use of the full range of the communication process, including feedback. For instance, the Shannon-Weaver model demonstrates communication as an interactive process where participants exchange messages and deal with noise in the channel. The next focus of our learning was analyzing the communication gap between FTF and online interactions with a high-touch approach of determining which method works best in certain situations. The knowledge was reinforced by reviewing the three major theories (EI, sensemaking, and CAT) and by analyzing interpersonal communication. The discussion then emphasized allowing communication participants to interact efficiently between the two methods regardless of the direction of movement. The discourse also emphasized employing verbal and non-verbal cues to interactions and helping online communicators find ways to apply some measure of those cues in the digital space. By covering a range of interpersonal and group communication theories, Chapter 7 energized the pursuit of bridging the gap in communication between FTF and online. All of the theories covered focus on how groups interact and manage conflict.

Section 3 moved our discussion to the high-teach emphasis of this book. A Social Networking Matrix provides basics of relevant theories along with two suggested approaches each for FTF and online interactions. An Interpersonal Communication Matrix explains relevant theories and offers reasons for interactions that are based on the theory, along with FTF and online suggested approaches for communicators. In comparing FTF and online communication, the next step in the journey was to offer the best ways to use each method with acknowledgment that there are positives and negatives in each. No approach will solve every communication issue, or answer every question, but we have discussed FTF and online in an attempt to be diverse in communicating. The tools to bridge the gap between face-to-face and online activities has now been provided. The lack of human contact in day-to-day communication is often negatively affected by social networks, with observed challenges of superficial communication experiences, degraded grammar and spelling, and language complications. The substance of face-to-face communication is intended to help senders and receivers get the most out of their interactions. So, this book characterizes FTF communication as basic, relevant, high-touch, powerful, complementary, interactive, timely, compelling, and committed (refer to Chapter 10 for specifics of these variables).

Hybrid Approach

The hybrid approach in Table 1 uses the best of our knowledge to identify approaches to communication that will work regardless of whether we're face-to-face or in a digital interaction. Crafting communication messages requires the sender's best skills, so we offer six steps to achieve success:

Table 1. Using a hybrid approach

Hybrid Approach
1. Start with Engaged Interaction (from Chapter 8) 2. Apply Interactionism 3. Apply Networked Individualism 4. Apply Communication Zones 5. Use the QWQ Formulas 6. Combine FTF and Online Interactions

Start with Engaged Interaction

These are three keys to how we employ flexible, full-range communications to ensure that both parties listen, hear, and understand, and continue the interaction until that happens. First, senders apply communication principles and then use the 4Rs of New Age Communications. The third key is employing strategic flexibility to measure influences in the environment and change the message for the best chance of success.

Apply Interactionism

Interactionism allows us to put together all the tools we've learned – SNIP, EVR, PIP, Engaged Interaction – to get the message across the right way every time, hopefully without fail. It truly allows us to listen, hear, and understand, regardless of whether our intended receiver is in the room or a world away. Jan Fernback of Temple University presented a view of interactionism that is perfectly suited for our comparison of FTF and digital conversations. In a 2007 article about a study of symbolic interactionist perspective on online social relations, Fernback suggests an evolving process that begins to address the issue of convenient togetherness without real responsibility (2007). Using interviews and a theoretical consideration of online community, the research effort found that some participants were anxious to discuss additional meaning that would be helpful in online social activity.

Our interactionist approach leverages the actions of social actors as we interpret the situations we face and make conscious choices on behavior. Social actors (senders) use the context of the interaction to adjust behavior to be complementary to the actions of other social actors (receivers). Interpretation of the situations we face affects how we encode and decode messages.

Apply Networked Individualism

This concept relates to the strategic flexibility in message delivery that is part of engaged interaction. Networked organizations have flexibility in messaging and actions, working with multiple and shifting work partners and execute partial involvements with shifting sets of workgroups. Ties extend across demographic boundaries and workers are free to choose their customers, the way in which they interact, and the time and place of those interactions (Wellman 2001). This is a social capital-building approach that focuses interaction on selected individual residents and their immediate social clusters (Wellman 2012). This peer-to-peer mode of exchange departs from a collectivist approach to community in "normal" mode of online communication, allowing online communicators some of the additional, personal information that is typically available only in FTF activities. Further, the smaller group interactions suggested have a more personal feel and usefulness, striving for fellowship, respect, and tolerance (Fernback 2007). An important aspect of networked individualism is that online communicators can associate with a church or a job or some other group to transcend physical place, thereby achieving continuity and commitment that is not available in the traditionally large community groups we see in the online environment.

Apply Communication Zones

This skill requires intimate social bonding, where the sender and receiver execute close ties that include regular feedback and have a "contractual" agreement to pay attention and respond to messages. While this two-pronged approach can work for either message, it is primarily an FTF skill that should be cultivated. The value of face-to-face communication cannot be ignored according to two researchers who argue that there is a rich body of research listing this as the most information-rich medium (Nardi and Whittaker 2002). The suggestion is that FTF persists in the workplace because it is the most effective way to establish and nurture human relationships that grow out of and support business relationships. Social bonding and symbolic expressions of commitment allow relationships to grow and trust to develop. The current argument is a contrast of social aspects of communication with information aspects such as information transfer, management of misunderstandings, the continuous search for common ground, and sharing information and objects in the environment (Nardi and Whittaker 2002). The information on FTF presented shows us that the most important consideration in communication is creating an environment in which people can interact freely regardless of the message that is being sent. Information exchange requires linkages between sender and receiver, and needs the participants to create and nurture feedback.

From reading the referenced article, it is clear that senders and receivers interact in what the authors call communication zones. These are social "fields" that support creativity and provide for productive communication between two people. Managing these zones in each interaction requires the creation of social ties or bonds, use and management of visual and verbal cues, and ongoing cooperation in feedback. Effective use of these zones involves two key processes. First, there are the social bonds mentioned here, because they allow people to connect emotionally. The second process is getting the parties to agree to "attentional contracts" that require that they pay attention to messages that are sent and received, and that they do sp for some undetermined amount of time (Nardi and Whittaker 2002). The researchers have focused their lens in what they consider a non-traditional way. In their view, most media and communication theories focus on interactions that are already underway. Their approach is to shine the light on how people prepare and engage in an environment where communication can take place. This is sort of like building a house in which to communicate instead of buying one that is already standing. This author agrees that the processes are important in achieving a state of interaction. The interaction should not be inserted into the environment, instead the environment should be constructed and nurtured to "receive" the sender and receiver and allow their interaction.

We've touched on the communication environment and included in that discussion is social bonding. Engaging sender and receiver in social interaction and use of informal conversation creates social bonding. Social bonding is about awareness, or having a sense of the other person's "body," which includes their physical appearance, body language, and facial expressions. Good communicators will also have a sense of the other person's makeup, jewelry, hair style, and clothing. Some of the research reviewed for this book indicated that people have a social, even complementary, perspective on the role of the body in communication. Social bonding benefits from the meaningful experiences between sender and receiver that can happen when they share a common physical space. Communicating in person creates benefits based on human contact, such as touching, eating, and drinking together. These interactions can be simulated in the digital conversation, or in some ways can be actual (via Skype or Face Time for instance). But only FTF can allow the full range of these social bonding benefits. Additional pluses come from off-the-cuff conversation such as family questions, gossip, jokes, or just idle chat. Social bonding is an invaluable consequence of FTF communication.

Use the QWQ Formula

This stands for "Question, W.A.I.T., Question," three formulas developed by Janine Driver in *"You Say More Than You Think"* that senders can use to find truth in communication (Driver and Aalst 2010). W.A.I.T. is an acronym for Why Am I Talking. It requires that the sender wait and carefully listen to what the receiver is saying before speaking again. In this approach, the sender asks pointed, specific questions to draw out information. The next step is to W.A.I.T. while the sender employs excellent listening skills to give the receiver more than ample time to respond. Once the response is received, the sender validates or seeks more information. The first set, Formula One, asks questions to assign the person a positive trait like trustworthiness. One receiver agrees and is finished responding, the sender validates the claim by seeking commonality, like saying "I'm trustworthy, too." This approach allows continued conversation and helps with relationship building. There are two other "formulas" as well, one that addresses information gathering in less confrontational negotiations, and one that detects deception and forces the receiver to confess. These are powerful tools that senders can use to engage receiver, share information, seek the truth, and create bonds that turn into relationships.

Combine FTF and Online Interactions

Online communications carry a potential risk of distrust, misunderstanding, and poor decision making if they are not connected with good interpersonal relationships. Being creative and adapting to personal and organizational needs and objectives can deliver unexpected benefits. "In some professional tasks CMC was found to be superior to FTF, but CMC was even more effective when combined with FTF." (Olaniran 1994) Senders should strive to improve their message-medium fit to optimize the choice of media and type of use. The enhanced collaboration that should result is focused on task, message, media-mix, interaction partners, CMC competence, context, and optimal outcomes (Bubaš 2001). FTF and online communicators can use the skills for enhanced collaboration to improve interactions.

The hybrid approach that is outlined here is rich in resources by which we can address the communication gap. Let's take a look at the intended benefits of taking this approach to communicating.

To begin, the hybrid approach offers a new way of communicating that adjusts to variable and diverse interactions. The approach is sensitive to the environment in that there are ways to assess internal and external forces and adjust the message accordingly. The formal approach should work to counter uncertainty, which was previously stated as a very serious challenge to communicating. Formality should result in building confidence in all communicators, allowing them to deal with new situations with or without assistance. Anthony Giddens discussed the importance of overcoming distance restrictions and allowing people to coordinate agendas in what he called a "pure relationship," which only lasts as long as it continues to offer all participants personal fulfillment. (Giddens 1984) This relationship is possible because actors have shared histories and can continue to build the social relationship.

Experts maintain that the mere presence of technology can and does inhibit developing interpersonal trust and closeness. When this happens, the individual seeks empathy and understanding from the interaction partner. Relevant research also suggests that extensive use of mobile phones hinders the formation of relationships. Larger, more active social networks tend to force face-to-face conversations out of the picture. Addressing the communication gap requires that participants adjust behaviors, attitudes, and cognitions to allow a new, more flexible approach to interactions.

FTF communication is very powerful in the way it can connect us. It's hard not to place ourselves in someone else's shoes when we experience emotion expressed on their face or read it in gesture or posture. The response happens quickly and automatically when face-to-face, because we are conditioned to react to things that affect us or come into our consciousness. We have a human need to match our position with ideas and attitudes that resonate with us because we want to feel what they feel and see what they see. This is the understanding and empathy that is so important to communicating. Social capital improves people's lives psychologically and biologically. Research suggests that close and frequent connections with others make us healthier.

Accepting the Communication Challenge: Closing the Gap

Senders achieve effective communication by identifying who they want to contact and by determining how the person or group wants to receive information. Senders can then create and maintain a two-way relationship and continuously seek feedback. Once the communication is alive and well, senders should continue to monitor the dialogue and follow some simple rules of engagement: listen, learn, respect, and share. All communication should be clear, creative, and collaborative, characterized by well-crafted messages, crafted in a way that is entertaining and compelling, and effective in getting both parties sending, receiving, and providing feedback on a regular basis. The collaborative approach is important in creating content that has value for all participants.

Social networking is wildly popular and extremely convenient. It's not hard to understand why someone would communicate online instead of face-to-face. However, there are still well-publicized barriers to digital communication that include fear, disbelief, underestimation of the benefits, lack of a sense that the time spent is valuable or necessary, misunderstanding of the capability, misperceptions about which platform to use (Facebook, Twitter, LinkedIn, Pinterest, etc.), or a combination of all of these reasons. You can navigate any or all of these barriers with an open mind, an understanding of personal social media evolution, and keys to engagement.

Communication is also about relationships and information sharing. Senders and receivers should agree on the purpose of the messaging activity and commit to an open and honest two-way exchange. Senders must try to understand the receiver's information needs, and attend to noise in the channel that includes emotions, relationships, and environment. Culture may also enter into the communication considerations, leading to the need to develop skills to handle diversity of thought and action between sender and receiver. Considering all of the variables that can affect the interaction provides an ability to ensure that the message is suitable for the current social context.

Communicators should not get discouraged if the receiver is not attentive, especially in terms of online communication. There is an online "1% participation rule." The rule states that for every 100 people online, only 1 person will create content and only 10 will "interact" with it. The other 89 will just view it. That means that most of what is sent results in no interaction. FTF activities would at least allow the sender to look other 89 receivers in the eye to seek a response.

Seeking Value

Every sender should be ultimately engaged in seeking value and in sharing that value with others. Shared information that is valuable can make interactions powerful and memorable. Senders and receivers should wish to be in open exchanges where constructive ideas and opinions are shared. They should enjoy active roles in the communication and take joint responsibility for listening, monitoring, and engaging in the conversation. These are results that both parties can enjoy and that are important to quality information sharing. Quality information sharing leads to building strong ties that can nurture long-term relationships.

Senders should establish their reputation as good communicators leading to valuable engagements that build social capital. Need satisfaction can be accomplished through collaboration where every partner is an equal contributor to the messaging activities. This should all be done hand-in-hand with analyzing the wants and needs of the receiver. Where does the receiver get their information and with whom do they normally exchange information? These are important questions that the sender should ask and answer. When conducting online communications, senders should try to determine where the receiver or target audience is posting, texting, viewing video, or sharing images.

So communication should be a series of action, response, and behavior in a two-way exchange. This is important regardless of the message. In this way, real value transcends time and space and leads to a continuity of connection that is priceless.

When communicating online, the conversation is structured in unconventional ways; it is organic and complex. It connects people and businesses, allowing engagements and shared content. This signals a valuable return on the investment of time. Even if there is simply a perception of benefit, users may be more inclined to participate, or they may be moved to increase their participation levels. Online senders should take advantage of this knowledge and deliver value-based information that improves the social networking and communication efforts.

When communicating face-to-face, the conversation must be dynamic. The process needs full attention to visual, verbal, and non-verbal cues for success. The shared information must have a shared value that all parties agree to. We are active communicators even when we don't want to be, so we should use and develop our skills at every opportunity. Great communication provides references to guide our thought processes, enhances positive perceptions, and addresses negative perceptions. Senders and receivers should cultivate a flexible interchange that fosters relationship building and leads to feedback. Feedback is crucial to make communication the cycle it should be.

Great engagements where people share valuable information and resources can enhance a person's performance or allow the digital "self" to be seen and heard beyond the internet. The benefits of conducting flexible communication do not stop there. The collaboration that comes from great interactions fosters coordination of efforts and management of tensions that can stifle communication and relationship building. In addition, effective communication can conquer some of the limitations of online interaction and can energize the face-to-face connection with the "human condition" that this book advocates as a solution to the communication gap. The focus of these interactions is shared value that creates ties that endure because of dependable communication channels, coordinated activities between participants, and the establishment of trusting relationships.

Regardless of the method, communicators should seek out people who share their values and engage. Vigilance is important to address interactivity with people you value. Then, continue to align values and trust to complement each other during the two-way communication. Engaging and delivering information coupled with properly addressing concerns and responsibilities can build enormous social capital. Listen and understand the feedback you receive and turn it into additional great messages. You will be rewarded with purposeful relationships to generate and receive tangible social, psychological, emotional, and economic benefits in short or long terms.

Overall, communicators can help themselves through interactive listening. The requirements for interactive listening are adapted for our purposes from a social media benchmarking study (Brown, Alkadry et al. 2013):

- Go beyond simply connecting with people and focus on building meaningful relationships that will endure;
- Develop a strong capability to listen to and understand all audiences;
- Trust and empower the communication partner(s) and strive for increased transparency in the interaction;
- Coordinate your messages and goals to make them clear and concise;
- Seek and return feedback at every opportunity;
- Be flexible to adapt to and respond to receivers in real time; and
- Seek and answer questions about the receiver's interest and satisfaction with the communication.

These are the things that communicators need to focus on. Senders need a good reputation as an information sharer and one who works to satisfy the receiver's needs. Senders should always seek shared value in the information they exchange. Communication, when done well, is about building social capital and lasting relationships, and about being flexible to enable sending the right message on the right platform at the right time.

The communication challenge that all senders must accept is to create interesting messages that spark interest and enthusiasm for the continued interaction, and that results in feedback. This is full-range communication at its best. Cultivate personal and group relationships as appropriate to get participants to think together and energize information sharing. Good communication transcends limitations of the technology through flexibility and creativity. A great personal relationship that allows people to engage in real dialogue builds social capital.

This has been a long communication journey to evaluate, dissect, and understand the communication gap that exists between face-to-face and online communication. We have examined theory, dispelled myths, addressed barriers, introduced new concepts, and turned each method upside down. The result should be a thorough understanding of the benefits and dangers, the pros and cons, and the ups and downs of each method. As you leave this communication discussion and go back to your daily life, I hope you have found plenty of value. I hope you have found something that brings you closer to the object of your communication effort, whether it is a person or a group. I hope that you have found something that makes messaging easy, fun, and interactive. Finally, I hope that you have found something that allows you to collaborate in every instance with great success.

REFERENCES

Brown, M. Sr, Alkadry, M., & Resnick-Luetke, S. (2013). Social Networking and Individual Perceptions: Examining Predictors of Participation. *Public Organization Review*, 1–20.

Bubaš, G. (2001). *Computer mediated communication theories and phenomena: Factors that influence collaboration over the Internet*. 3rd CARNet Users Conference, Zagreb, Hungary.

Driver, J., & Aalst, M. v. (2010). *You say more than you think: the 7-day plan for using the new body language to get what you want*. New York: Crown Publishers.

Fernback, J. (2007). Beyond the diluted community concept: A symbolic interactionist perspective on online social relations. *New Media & Society*, *9*(1), 49–69. doi:10.1177/1461444807072417

Giddens, A. (1984). *The constitution of society: Outline of the theory of structuration*. Univ of California Press.

Nardi, B. A., & Whittaker, S. (2002). The place of face-to-face communication in distributed work. *Distributed Work*, 83-110.

Olaniran, B. A. (1994). Group performance in computer-mediated and face-to-face communication media. *Management Communication Quarterly*, *7*(3), 256–281. doi:10.1177/0893318994007003002

Wellman, B. (2001). Little boxes, glocalization, and networked individualism. In Digital cities II: Computational and sociological approaches. Springer.

Wellman, B. (2012). Networked individualism: how the personalized internet, ubiquitous connectivity, and the turn to social networks can affect learning analytics.*Proceedings of the 2nd International Conference on Learning Analytics and Knowledge*. ACM. doi:10.1145/2330601.2330603

About the Author

Michael A. Brown, Sr., Ph.D., earned his Public Administration and Urban Policy degree, International Business, from Old Dominion University (ODU) in May 2011. He is teaching online social media, public relations and communication courses for Florida International University (FIU). Three of his online courses at FIU are recognized for excellence in education by Quality Matters (QM). QM is a nationally recognized faculty peer review organization for online and hybrid course design, signaling the best offerings in education. He is an Air Force civil servant working as the deputy director of Public Affairs for a joint military organization. This PR professional has 40-plus years of military and civilian experience combined, and is an Air Force retiree who served 24 years in uniform.

Index

www.ingramcontent.com/pod-product-compliance
Lightning Source LLC
LaVergne TN
LVHW080853300926
849162LV00003B/416